ODC 42 FOR

£40

IEWTON RIGG COLL

Telephone 01768 863791

or renewed before
below

IUFRO Research Series

The International Union of Forestry Research Organizations (IUFRO), with its 14,000 scientists from 700 member institutions in 100 countries, is organized into nearly 300 research units that annually hold approximately 60 conferences, workshops and other meetings. The individual papers, proceedings and other material arising from these units and meetings are often published but in a wide array of different journals and other publications. The object of the IUFRO Research Series is to offer a single, uniform outlet for high quality publications arising from major IUFRO meetings and other products of IUFRO's research units.

The editing, publishing and dissemination experience of CABI *Publishing* and the huge spread of scientific endeavours of IUFRO combine here to make information widely available that is of value to policy makers, resource managers, peer scientists and educators. The Executive Board of IUFRO forms the Editorial Advisory Board for the Series and provides the monitoring and uniformity that such a high quality series requires in addition to the editorial work of conference organizers.

While adding a new body of information to the plethora currently dealing with forestry and related resources, this Series seeks to provide a single, uniform forum and style that all forest scientists will turn to first as an outlet for their conference material and other products, and that the users of information will also see as a reliable and reputable source.

Although the official languages of IUFRO include English, French, German and Spanish, the majority of modern scientific papers are published in English. In this series, all books will be published in English as the main language, allowing papers occasionally to be in other languages. Guidelines

for submitting and publishing material in this series are available from the
Publisher, Books and Reference Works, CABI Publishing, CAB International,
Wallingford, Oxon OX10 8DE, UK and the IUFRO Secretariat, c/o Federal
Forest Research Centre, Seckendorff-Gudent-Weg 8, A-1131, Vienna, Austria.

Forest Dynamics in Heavily Polluted Regions
Report No. 1 of the IUFRO Task Force on Environmental Change

Edited by

J.L. Innes
Forest Resources Management, University of British Columbia, Vancouver, Canada

and

J. Oleksyn
Polish Academy of Sciences, Institute of Dendrology, Kórnik, Poland

CABI *Publishing*
in association with
The International Union of Forestry Research Organizations (IUFRO)

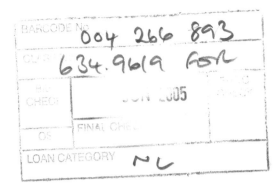
CABI *Publishing* is a division of CAB *International*

CABI Publishing
CAB International
Wallingford
Oxon OX10 8DE
UK

Tel: +44 (0)1491 832111
Fax: +44 (0)1491 833508
Email: cabi@cabi.org

CABI Publishing
10E 40th Street
Suite 3203
New York, NY 10016
USA

Tel: +1 212 481 7018
Fax: +1 212 686 7993
Email: cabi-nao@cabi.org

Published in Association with:

The International Union of Forestry Research Organizations (IUFRO)
c/o Federal Forest Research Centre
Seckendorff-Gudent-Weg 8
A-1131 Vienna
Austria

ISBN 0 85199 376 1

Typeset by AMA DataSet Ltd, UK
Printed and bound in the UK by Biddles Ltd, Guildford and King's Lynn

Contents

Contributors

M.J. Arbaugh, Pacific Southwest Research Station, Forest Service, USDA, 4955 Canyon Crest Drive, Riverside, CA 92507, USA

K. Bull, ITE Monks Wood, Abbots Ripton, Huntingdon PE17 2LS, UK

V. Caboun, Forestry Research Institute (LVU), Zvolen, Slovakia

G.A. Evdokimova, The Institute of the North Industrial Ecology Problems, Kola Science Centre, Russian Academy of Science, Fersmana Street 14, 184 200 Apatity, Russian Federation

G. Fenech, Environment Canada, 4905 Dufferin Street, Downsview, Ontario M3H 5T4, Canada

B. Freedman, Department of Biology, Dalhousie University, Halifax, Nova Scotia B3H 4J1, Canada

A. Gorzelak, Forest Research Institute (IBL), Warsaw, Poland

M.L. Gytarsky, Institute of Global Climate and Ecology, 20-B Glebovskaya St, Moscow, Russian Federation

E. Haukioja, Section of Ecology, Department of Biology, University of Turku, FIN-20014 Turku, Finland

J.L. Innes, Forest Resources Management, University of British Columbia, 2045, 2424 Main Mall, Vancouver, British Columbia V6T 1Z4, Canada

R.T. Karaban, Institute of Global Climate and Ecology, 20-B Glebovskaya St, Moscow, Russian Federation

V.I. Kharuk, Sukachev Institute of Forest, 660036 Krasnoyarsk, Russian Federation

M.V. Kozlov, Section of Ecology, Department of Biology, University of Turku, FIN-20014 Turku, Finland

M. Kytö, Vantaa Research Centre, Finnish Forest Research Institute, PO Box 18, FIN 01301 Vantaa, Finland

A. Luttermann, Department of Biology, Dalhousie University, Halifax, Nova Scotia B3H 4J1, Canada

M. Mikulowski, Forest Research Institute (IBL), Warsaw, Poland

M.M. Millán, Centre for Environmental Studies of the Mediterranean, Parque Tecnologico, Calle 4 - Sector Oeste, 46980 Valencia, Spain

P.R. Miller, Pacific Southwest Research Station, Forest Service, USDA, 4955 Canyon Crest Drive, Riverside, CA 92507, USA

P. Niemelä, Vantaa Research Centre, Finnish Forest Research Institute, PO Box 18, FIN 01301 Vantaa, Finland

J. Oleksyn, Polish Academy of Sciences, Institute of Dendrology, Parkowa 5, PL-62-035 Kórnik, Poland

G. Raben, Saxon State Institute for Forestry (LAF), Graupa, Germany

O. Rigina, Institute of Geography, University of Copenhagen, Copenhagen, Denmark

M.-J. Sanz, Centre for Environmental Studies of the Mediterranean, Parque Tecnologico, Calle 4 - Sector Oeste, 46980 Valencia, Spain

K. Vancura, Branch of Forest Management, Ministry of Agriculture of the Czech Republic, Tesnov 17, 11705 Prague 1, Czech Republic

N.P. Vassilieva, All-Russian Institute for Nature Conservation, Znamenskoe-Sadki, Moscow, Russian Federation

K. Winterhalder, Department of Biology, Laurentian University, Sudbury, Ontario, P3E 2C6, Canada

E. Zvereva, Section of Ecology, Department of Biology, University of Turku, FIN-20014 Turku, Finland

Foreword

Major changes are occurring in the way that science is being undertaken. Until now, there has been a tendency for scientists to work in isolation, looking at very specific problems. However, many policy issues cut across a number of scientific disciplines, and the impact of environmental change on forests is one such issue. It was for this reason that IUFRO established a Task Force on Environmental Change in 1995. This report is the first product of the Task Force, and illustrates well the international, interdisciplinary nature of scientists involved in IUFRO.

One of the features of air pollution control has been the development of international agreements such as the Convention on Long-Range Transboundary Air Pollution. These agreements have laid the basis for cuts in pollutants known to be damaging the environment. The development of such agreements requires high-quality scientific knowledge about the nature and extent of the impacts of specific pollutants. This is especially true given the recent change in emphasis away from across-the-board agreements (such as a 30% cut in national emissions of sulphur) to effects-based reductions, which rely on knowledge of the likely impacts of a pollutant on specific sensitive receptors. IUFRO has played a major part in providing such information, initially through Project Group P2.05 and its several Working Parties and latterly through the new Division 7. Air pollution research was one of the earliest cooperative efforts within IUFRO, with trials being set up in the 'Black Triangle' of the former East Germany, Poland and Czechoslovakia in the 1940s and 1950s. It is therefore highly appropriate that one of the editors of this report is a Polish geneticist working on pollution problems.

I very much welcome the publication of this report, which should provide scientists throughout the world with a state-of-the-art assessment of the extent of air pollution impacts in heavily polluted regions. The associated summary for policy makers is also welcome, and reflects the increasing role that IUFRO intends to play in ensuring that policy makers are fully informed of scientific issues and that they are adequately informed of the consequences of their decisions.

Professor Jeff Burley
President, International Union of Forestry Research Organizations

Acknowledgements

We are grateful to the Editorial Board for this volume (David Karnosky, Niels Elers Koch and Lisa Sennerby-Forsse) for their constructive comments on the structure of the report. Many people were involved in the review process, and we thank them for their helpful comments on individual chapters. Mrs Vreni Fataar of the Swiss Federal Institute for Forest, Snow and Landscape Research kindly prepared the diagrams.

Air Pollution and Forests in Heavily Industrialized Regions: an Introduction

J. Oleksyn[1] and J.L. Innes[2]

[1]*Polish Academy of Sciences, Institute of Dendrology, Kórnik, Poland;* [2]*Department of Forest Resources Management, University of British Columbia, Vancouver, Canada*

Air pollution represents a threat to many forests, particularly in the Russian Federation. Various types of air pollution affect forests, with the most significant pollutants being sulphur dioxide, fluorides, heavy metals and ozone. In the past, these have been responsible for the destruction of substantial areas of forest. In North America and Western Europe, such effects have now been brought under control, and steps are being taken to aid the recovery of damaged ecosystems. Eastern Europe remains in a transition stage, with only a poor understanding of the complex factors causing the declines of specific forests. In the Russian Federation, the pollution in some areas remains at very high levels, although economic developments in the 1990s have resulted in lower pollution emissions associated with lower production. However, the damage in some areas of the country continues to expand, with more than 1 million hectares in the vicinity of Noril'sk, Siberia, showing visible signs of damage, and an estimated 7 million hectares affected. International initiatives to control pollution on a multilateral or bilateral basis have resulted in significant improvements. Similar initiatives by the research community could result in major benefits.

1 Introduction

For almost 150 years, air pollution has been recognized as being a potential problem for forests. Early industrialization was associated with the large-scale cutting of timber for fuel, and many currently deforested areas around smelters (e.g. at Sudbury, Canada, see Winterhalder, this volume) can be attributed to felling rather than to the direct impacts of air pollution on trees. The demand for wood as a fuel for ore smelting resulted in the conversion of many natural forests to coppice systems. This is still evident in many forests of the northern hemisphere, with former coppice gradually reverting to high forest following either the substitution of wood by electricity or oil, or by the abandonment of small-scale furnaces following the exhaustion of local ores or the consolidation of the furnaces into large-scale smelting operations.

Today, a number of different types of pollution are associated with industrial and urban areas (see Chapter 2, this volume). Small-scale industries remain important in some areas, but it is usually large-scale industries that are the most significant sources of sulphur dioxide and heavy metals. This has been recognized in Europe and North America, with the European Community having a Large Combustion Plant Directive specifically aimed at combating such problems. Less rigorous controls exist in Eastern Europe and places such as the Kola Peninsula of the Russian Federation, and high levels of pollutants such as sulphur dioxide, fluorides and heavy metals remain a significant problem. At present, the most important factor affecting emissions in such areas is the level of production. Economic declines over the last 10 years are leading to substantial reductions in emissions associated with lower rates of industrial production (Chapter 4, this volume; Šrámek, 1998).

Many areas in Western Europe and North America have been severely affected by heavy metals and gaseous pollutants in the last 150 years. Sites such as Trail, British Columbia (McBride, 1937), and Sudbury, Ontario (Gorham and Gordon, 1960; Gunn, 1995), in Canada, Copper Hills in Tennessee, USA (Haywood, 1910; Hedgecock, 1914; Hursh, 1948), the Ruhr and other parts of Germany (Stöckhardt, 1853; Schröter, 1907), and the Black Country in England (Farrar et al., 1977) are all renowned for the environmental problems associated with industrial activities. Many smaller areas have also been badly affected by particular industries. For example, fluorides from aluminium smelting created major problems in areas such as Valais, Switzerland (Flühler, 1981, 1983; Flühler et al., 1981), the Arc Valley, France (Tessier et al., 1990), Øvre Årdal, Norway (Braanas, 1970), Spokane-Mead, Washington, USA (Adams et al., 1952) and Kitimat, British Columbia, Canada (Bunce, 1989).

In the late 1970s and early 1980s, there was considerable concern that the problems experienced in the heavily industrialized parts of Poland, and the former East Germany and Czechoslovakia were also occurring in more remote areas. However, the predictions of a massive and permanent loss of forests over much of Europe did not materialize (Skelly and Innes, 1994; Kandler and Innes, 1995), although a number of pollution-related problems were identified in specific areas. Today, despite periodic reports to the contrary in the press, the available information does not suggest that any major decline of forests in Western Europe is imminent.

Increasingly, it is not sulphur dioxide and heavy metals that represent problems, but ozone. Ozone is a secondary pollutant, derived from reactions between nitrogen dioxide and volatile organic compounds (VOCs). The primary source for the nitrogen dioxide is often automobiles, making it much more difficult to control. The classic area for such problems is California (Chapter 9, this volume), but other areas such as the Mediterranean region in Europe may also be affected (Chapter 10, this volume). This particular problem is likely to increase in significance in the future.

2   Contents of this book

The chapters of this book are arranged in what we believe to be a logical order. Luttermann and Freedman (Chapter 2) provide background information on different types of air pollution for the non-specialist, describing the main types of pollutants found in heavily industrialized regions. The Russian Federation is responsible for the management of a significant part (22%) of the world's forest resources; however, substantial areas of these forests are under threat from pollution, and Vassilieva, Gytarsky and Karaban (Chapter 3) provide an idea of the extent of the problem and how it is being monitored. A severely affected area is the Kola Peninsula, in the north-western part of the country (Fig. 1.1) and some of the problems here are described by Rigina and Kozlov (Chapter 4).

Fig. 1.1.   The smelter at Nikel, on the Russian–Norwegian border. See Chapter 4 for a detailed description of this area.

The severity of the impacts in this area provide the opportunity to determine the effects of industrial pollution on a range of ecosystem components that are normally more difficult to assess than trees. Evdokimova provides a summary of the effects on the microbial diversity of the soils (Chapter 5). Soil microbes play a critical part in the functioning of forest ecosystems, so this review is important and relevant. Another area in the Russian Federation severely affected by air pollution and described by Kharuk (Chapter 6) is the surroundings of Noril'sk, in Siberia. Here, as much as 7 million hectares (Mha) of forest may be affected, with damage being visible on more than 1 Mha (Sokolov, 1997). Faced with such statistics, the huge sums of money spent on researching purported air pollution problems in Western Europe and eastern North America in the 1980s suggest that global priorities for the investigation of forest problems require better coordination.

The examples from the Russian Federation are depressing. However, similar problems have been experienced in other parts of the world, and have, to a lesser or greater extent, been resolved. Winterhalder (Chapter 7) provides an account of the environmental degradation experienced at Sudbury, Canada, which has one of the two most important nickel sulphide deposits in the world (the other is at Noril'sk). The photographs that he has brought together for this chapter provide a vivid impression of the way in which research can help resolve a very real problem. However, the recovery would have been much less apparent if the local community had not been involved, and it is appropriate that the Sudbury community received the 1992 United Nations Local Government Honours Award for its work to reverse environmental degradation in the area. Eastern Europe provides an intermediate stage between the problems of the Russian Federation and those of Western Europe and North America. With large-scale investments currently occurring in the area, the pollution problems described by Vancura, Raben, Gorzelak, Mikulowski, Caboun and Oleksyn (Chapter 8) should be of historical interest only. However, this is not the case, and many forests in the area are still experiencing problems. The role of pollution is uncertain; poor matching of provenance or even species to site, and the use of Norway spruce (*Picea abies* (L.) Karst.) monocultures, are now also believed to have played a role (e.g. Kulhavý and Klimo, 1998; Šrámek, 1998).

The majority of classic sites of air pollution injury to forests are caused by sulphur dioxide, fluorides, heavy metals or combinations of these pollutants. However, ozone is of increasing concern. Most people have heard of the Los Angeles smogs, happily now not nearly as extreme as in the past. Ozone was a major component of the smog, and is still present in California in sufficient quantities to cause injury to trees in remote areas. Miller and Arbaugh (Chapter 9) describe the extent of the problem. Characteristically, there is no reference to the fact that it was the work of Paul Miller and his colleagues in the 1960s that identified ozone as the agent responsible for the mysterious 'X-disease' affecting ponderosa pine in the mountains due east of Los Angeles and previously attributed to a wide range of other factors. This work is one of the classic studies involving the establishment of a cause–effect relationship between forest health and air pollution, and we are pleased to note that the material has recently been summarized in a monograph (Miller and McBride, 1999). Ozone is increasingly a problem in other parts of the world, and Sanz and Millán (Chapter 10) describe the nature of ozone dynamics and its effects on forests in the eastern Mediterranean.

Various steps can be taken to ameliorate the impacts of pollution. This is a huge area, and it has only been possible to touch on it in this book. Kozlov, Haukioja, Niemelä, Zvereva and Kytö (Chapter 11) provide a brief overview of some of the approaches that have been adopted, as does Winterhalder in his contribution (Chapter 7). Gunn (1995) provides a more detailed account of the greening up of the Sudbury area. The use of forestry in the reclamation of degraded areas is one that is extremely topical but it was considered outside the scope of the Task Force on Environmental Change. However, we strongly recommend that it receives more attention from the scientists of the International Union of Forest Research Organizations.

Bull and Fenech (Chapter 12) provide grounds for hoping that, in the long term, pollution problems will be solved soon after their identification. The Convention on Long-Range Transboundary Air Pollution has provided a forum for governments with the United Nations Economic Commission for Europe to meet and discuss ways in which transboundary pollution can be controlled. As Bull and Fenech describe, there are also other bodies that are active, whether it is the bilateral arrangements between Canada and the USA or the concerted efforts of the member states of the European Union. A particularly important point they make is that air pollution controls are increasingly being based on the expected benefits. These are no longer restricted to human health effects. Damage to forests is also seen as important, although it has received much less attention than it deserves. With the growing realization that forests are an essential part of the human ecosystem, perhaps we shall see greater concern for forest effects in future.

3   Future developments

In many areas, pollutant concentrations have decreased over the last 10 years, not because of the installation of control technologies, but because of reductions in production. Consequently, the improvements in air quality should not be seen as permanent. Economic conditions may change, and this needs to be considered when looking at trends in pollutant concentrations in such areas. The Kola Peninsula in the Russian Federation is an example; much of the sulphur dioxide that is emitted is the result of the use of high-sulphur ores imported from Noril'sk. As long as it remains economically feasible to ship the high-sulphur nickel sulphide ores from Noril'sk to the Kola Peninsula, pollution problems will continue.

New smelters will continue to be constructed. For example, large new aluminium smelters are currently being built at Alma, Quebec, and Maputo in Mozambique, at a combined cost of almost US$3 billion. With current technology, it is possible to ensure that these produce minimum levels of pollution. However, it is worth noting that a substantial proportion of the power for the Mozal smelter in Mozambique will come from electricity generated by burning brown coal in South Africa. Existing sources of pollution are more difficult to control. In some cases, the local ore bodies may become exhausted, but the investment in plant is such that ores from elsewhere may be imported, as at Zapolyarnyy. A variety of possible mechanisms exists for reducing the pollution from such sources, and the main problems seem to be related more to financial resources and politics than technological difficulties (cf. the problems described by Rigina and Kozlov, this volume, at Pechenganikel). Reducing pollution from small, diffuse sources, such as cars, may prove more difficult to control. Ozone represents a growing problem in many areas, and exceedances of critical levels are widespread. The resolution of this issue will be a major challenge for the future.

One of the requirements for sustainable forest management is that forests are protected from the damaging effects of air pollution (Montreal Process, Criterion 3). Foresters themselves are not in a position to control air pollutants. However, they can and should draw attention to any adverse effects that are identified in the forest. This book should provide an indication of the sort of effects that can be anticipated.

References

Adams, D.F., Mayhew, D.J., Gnagy, R.M., Richey, E.P., Koppe, R.K. and Allan, I.W. (1952) Atmospheric pollution in the ponderosa pine blight area, Spokane County, Washington. *Industrial and Engineering Chemistry* 44, 1356–1365.

Braanas, T. (1970) Fluorskader på furu i Vettismorki. *Tidsskrift for Skogbruk* 78, 379–401.

Bunce, H.W.F. (1989) The continuing effects of aluminium smelter emissions on coniferous forest growth. *Forestry* 62, 223–231.

Farrar, J.F., Relton, J. and Rutter, A.J. (1977) Sulphur dioxide and the scarcity of *Pinus sylvestris* in the industrial Pennines. *Environmental Pollution* 14, 63–68.

Flühler, H. (1981) Eine Übersicht über die laufenden Untersuchungen und bisherigen Ergebnisse. *Mitteilungen, Eidgenössische Anstalt für das forstliche Versuchswesen, Birmensdorf* 57, 367–372.

Flühler, H. (1983) Longtermed fluoride pollution of a forest ecosystem: time, the dimension of pitfalls and limitations. In: Ulrich, B. and Pankrath, J. (eds) *Effects of Accumulation of Air Pollutants in Forest Ecosystems.* D. Riedel, Dordrecht, pp. 303–317.

Flühler, H., Keller, T. and Schwager, H. (1981) Die Immissionsbelastung der Föhrenwälder im Walliser Rhonetal. *Mitteilungen, Eidgenössische Anstalt für das forstliche Versuchswesen, Birmensdorf* 57, 399–414.

Gorham, E. and Gordon, A.G. (1960) Some effects of smelter pollution northeast of Falconbridge, Ontario. *Canadian Journal of Botany* 38, 477–487.

Gunn, J.M. (ed.) (1995) *Restoration and Recovery of an Industrial Region. Progress in Restoring the Smelter-Damaged Landscape near Sudbury, Canada.* Springer-Verlag, New York.

Haywood, J.K. (1910) *Injury to Vegetation and Animal Life by Smelter Wastes.* Bulletin No. 113. US Department of Agriculture, Bureau of Chemistry, Washington, DC.

Hedgecock, G.G. (1914) Injuries by smelter smoke in southeastern Tennessee. *Journal of the Washington Academy of Science* 4, 70–71.

Hursh, C.R. (1948) *Local Climate in the Copper Basin of Tennessee as Modified by the Removal of Vegetation.* Circular No. 774. US Department of Agriculture, Washington, DC.

Kandler, O. and Innes, J.L. (1995) Air pollution and forest decline in Central Europe. *Environmental Pollution* 90, 171–180.

Kulhavý, J. and Klimo, E. (1998) Soil and nutrition status of forest stands under various site conditions of the Moravian-Silesian Beskids. *Chemosphere* 36, 1113–1118.

McBride, C.F. (1937) *Rossland Region Survey and Preliminary Management Plan.* Forest Surveys and Working Plans Division, BC Forest Service, Victoria, British Columbia.

Miller, P.R. and McBride, J.R. (eds) (1999) *Oxidant Air Pollution Impacts in the Montane Forests of Southern California. A Case Study of the San Bernardino Mountains,* Ecological Studies 134. Springer-Verlag, New York.

Schröter, G. (1907) Die Rauchquellen im Königreich Sachsen und ihr Einfluß auf die Forstwirtschaft. *Tharandter Forstliches Jahrbuch* 57, 211–430.

Skelly, J.M. and Innes, J.L. (1994) Waldsterben in the forests of Central Europe and eastern North America: fantasy or reality? *Plant Disease* 78, 1021–1032.

Sokolov, P.A. (1997) Ecological review of some problems in the sphere of forest use within the Russian Federation. In: Anttonen, T. and Pavlovich, A. (eds) *Potential of the Russian Forests and Forest Industries.* Research Notes 61. University of Joensuu, Joensuu, Finland, pp. 51–56.

Šrámek, V. (1998) SO_2 air pollution and forest health status in northwestern Czech Republic. *Chemosphere* 36, 1067–1072.

Stöckhardt, A. (1853) Untersuchungen junger Fichten und Kiefern, welche durch den Rauch der Antonshütte krank geworden. *Tharandter forstliches Jahrbuch* 9, 169–172.

Tessier, L., Serre-Bachet, F. and Guiot, J. (1990) Pollution fluorée et croissance radiale des conifères en Maurienne (Savoie, France). *Annales des Sciences Forestières* 47, 309–323.

Risks to Forests in Heavily Polluted Regions

2

A. Luttermann and B. Freedman

Department of Biology, Dalhousie University, Halifax, Canada

Human populations and industrial activities have increased dramatically during the past several centuries. This has resulted in enormous emissions of air pollutants in many regions of the world, which continue to affect surrounding forests and other ecosystems. Here, we describe the major constituents of air pollution typically found in heavily industrialized and urbanized regions, and the risks that these pose to nearby forests. From the perspective of forest health, the most important gaseous pollutants are sulphur dioxide (SO_2), ozone (O_3), hydrogen sulphide (H_2S), oxides of nitrogen (NO_x), ammonia (NH_3) and fluorides (especially HF). Other air pollutants which can have considerable effects include heavy metals suspended in gaseous emissions, such as copper and nickel, and the acidifying constituents of dry and wet deposition. Substantial variations in the effects of these pollutants are observed, ranging from acute local impacts to regional low-level changes in ecosystem composition and functioning. A commonly observed spatial pattern is a geometrical decrease in the intensity of effects spreading over a region away from a point-source.

1   Introduction

From an ecological perspective, air pollution may be defined as a condition in which concentrations of substances in the air are elevated above typical background levels to the degree that they cause measurable and undesirable effects on organisms and/or ecosystems (Freedman, 1995; Seinfeld and

Pandis, 1998). Air pollutants may originate from natural sources, such as volcanic activity and forest fires, as well as from large-scale anthropogenic emissions. Examples of the latter include industrial processes such as metal smelting and refining, the production of commercial energy through the burning of fossil fuels, and the use of fossil-fuelled engines for transportation. Anthropogenic emissions of air pollutants have increased enormously during the past century, due to a combination of population growth and intensified industrialization. As a result, many regions of the world experience high concentrations of air pollutants, which in some cases have caused significant ecological effects, including damage to forests (Miller and McBride, 1975; Heck and Brandt, 1977; Nriagu, 1984; Treshow, 1984; Legge and Krupa, 1986; Shriner, 1990; Smith, 1990; Kato and Akimoto, 1992; Freedman, 1995; Seinfeld and Pandis, 1998). This chapter defines the major constituents of air pollution typically found in heavily industrialized and urbanized regions, and describes in general terms the ecological risks these pose to forests. Subsequent chapters examine detailed case studies of specific effects that have been observed in various places and regions.

The risks to forest ecosystems caused by air pollution are dependent upon numerous variables. The most important of these are: the short- and longer-term concentrations of pollutants, frequency of exposure, topography, soil types, climate, seasonal weather patterns, species composition of vegetation (including the sensitivity of prominent species to chemical stressors), age structure of prominent species, plant phenology and the cumulative effects of other environmental stressors (Legge and Krupa, 1986; Darrall, 1989; National Research Council, 1989). Plant species vary considerably in their tolerance of various pollutants (Singh *et al.*, 1991). Due to the extreme complexity of exposure scenarios and biological/ecological vulnerability to air pollutants, it can be difficult to generalize about the specific effects of air pollution in any particular region. However, in cases involving exposures to high concentrations of pollutants emitted by discrete point-sources, it is often observed that damage is most severe close to the point-source, and that it decreases in a more or less geometric fashion with increasing distance away from the source (see for example the contributions by Kharuk, Rigina and Kozlov, and Winterhalder, this volume).

The effects of environmental stressors on a forest ecosystem can be studied at various levels. The focus may be on changes in the physiology of individuals of certain species, or more broadly, on their population dynamics. Community-level attributes, such as competition, productivity and nutrient cycling, can also be investigated. At the most extensive scale, landscape-level elements of the structure and function of ecosystems, such as hydrology, biogeochemical cycling, carbon storage and patch-dynamic succession can be the focus of attention (Smith, 1990). Currently, the short-term responses of individuals or populations of particular species (usually highly valued or indicator species) exposed to specific pollutants are better understood than effects at community or landscape levels. Even less is known about the potential ecological responses to combinations of pollutants over the long term, particularly at the level of forest landscapes. Broadly observed effects resulting from environmental stresses on forests are summarized in Table 2.1.

Direct cause–effect relationships can be difficult to establish, as many stressors are likely to be acting simultaneously (Darrall, 1989). In addition, observable effects may be temporally dissociated from the period when exposure occurred. For example, Strand (1993) found that summer exposure to SO_2 and NO_2 affected the photosynthetic activity of Scots pine (*Pinus sylvestris* L.) in winter. In some cases, however, a particular stressor may have an obvious primary effect on the health of a forest ecosystem, as in the event of acute damage caused by severe air pollution.

The geographical scale of interest also influences any discussion of the potential effects of air pollutants on forests. For example, the effects of regional air pollution on forests may be widespread and subtle, as in the case of acidification potentially associated with the long-range transport of acidifying

Table 2.1. Effects of environmental stress on forest vegetation. (After Darrall, 1989; National Research Council, 1989; Wolfenden and Mansfield, 1991.)

Effects on individual trees	Effects on forests
Visible symptoms of injury include changes in: • shape, size, colour and number of leaves • patterns of growth and development • patterns of foliage growth and senescence	Decreased productivity and living biomass of stands
Alterations in physiological processes, such as: • photosynthesis • respiration and metabolism • transpiration • mineral nutrition • stomatal function • transport and allocation of photosynthate • carbon allocation • hormonal control of growth • symbiotic relationships with other organisms	Changes in: • age-class distribution • patterns of competition and mortality • patterns of community succession • species composition • nutrient cycling • hydrological cycles • genetic structure of populations • biomass and carbon storage
Changes in susceptibility to other environmental stressors	
Life-history changes, such as: • decreased longevity; early onset of senescence • alterations in reproductive behaviour	

substances in the atmosphere (this relates to both wet and dry deposition). Conversely, in a heavily industrialized area, the effects of high levels of atmospheric pollution may be demonstrated quite visibly in surrounding forests and other ecosystems (e.g. Nriagu, 1984; Pryde, 1991; Peterson, 1993; Freedman, 1995; Gunn, 1995; Hansen *et al.*, 1996). In such cases of intense air pollution, vegetation may suffer extensive morbidity or mortality, resulting in dramatic changes in species composition, reductions in biomass and productivity, erosion of soil, nutrient loss, and changes in microclimatic and hydrological regimes. In severe cases, the perturbation can result in extreme ecosystem simplification and almost complete denudation (Bormann, 1982; Smith, 1990; Freedman, 1995). Some of the most intense cases of anthropogenic air pollution have occurred in temperate and boreal forest ecosystems, although an increasing number of cases are being documented from the tropical zone.

2   Major constituents of air pollution affecting forests

From the perspective of forest health, the most important gaseous pollutants are sulphur dioxide (SO_2), ozone (O_3), hydrogen sulphide (H_2S), oxides of nitrogen (NO_x), ammonia (NH_3) and fluorides (especially HF). Typical concentrations of these substances in relatively clean and polluted air are given in Table 2.2. Other constituents that can have considerable effects include heavy-metal particulates suspended in gaseous emissions, such as copper, lead and nickel. The wet and dry deposition of acidifying substances from the

Table 2.2.   Typical mixing ratios of important gaseous air pollutants in the troposphere (ppb). For the purpose of the discussion in this chapter, general values are considered sufficient for comparison. The units used are mixing ratios by volume: the mass of a substance as a mole fraction of all gaseous substances in a given volume of air. Although these values will vary with humidity (as water vapour is included in the total), other measures of concentration, e.g. mol m^{-3} also vary according to pressure and temperature. Seinfeld and Pandis (1998) provide a detailed discussion of units and conversions commonly employed in atmospheric chemistry. (Sources: Smith, 1990; Freedman, 1995; Seinfeld and Pandis, 1998.)

Gas	Relatively clean air	Polluted air	Atmospheric residence time
Sulphur dioxide (SO_2)	0.2–10	20–200	4 days
Hydrogen sulphide (H_2S)	0.2	nd	1–3 days
Nitric oxide (NO)	0.2–2	50–750 ⎫	
Nitrogen dioxide (NO_2)	0.5–4	50–250 ⎬	NO_x 1–4 days
Ozone (O_3)	10–80	100–500	
Ammonia (NH_3)	0.1–1	10–25	7 days
Fluorides	0.3	3–100	

atmosphere can also be important, particularly the dry deposition of SO_2, NO_x and NH_3 in areas where these occur in high concentrations at ground level.

In and around densely populated urban areas, the most important air pollutants are SO_2 and O_3. Urban areas emit large quantities of the precursors of acidifying deposition, which may be more elevated in remote regions. Fluorides, ethylene, NO_x, NH_3, chlorine, hydrogen chloride and particulates are also components of air pollution in urban areas, although they rarely occur in concentrations high enough to cause demonstrable damage to vegetation (Pignata *et al.*, 1997). Damage to forest vegetation in the immediate vicinity of geothermal plants has also been documented. For example, in a study reported by Bussotti *et al.* (1997), exposure to boron and arsenic was determined to be the likely cause of the observed damage. For the purposes of the present discussion, we will focus on some of the risks associated with SO_2, nitrogen gases, O_3, fluorides, heavy metals and acidic deposition.

2.1   Sulphur dioxide

Anthropogenic emissions of gaseous sulphur occur primarily as SO_2. Hydrogen sulphide (H_2S) forms a smaller proportion of the emissions, and is converted rapidly into SO_2 in the atmosphere (the residence time is typically < 1–2.9 days) (Berresheim *et al.*, 1995; Seinfeld and Pandis, 1998). SO_2 is eventually oxidized to the anion SO_4^{2-} (sulphate) in the atmosphere, or after the SO_2 has dry-deposited to a terrestrial or aquatic surface. Factors influencing the rate of oxidation include the levels of sunlight, humidity, and the presence of nitrogen oxides, hydrocarbons, strong oxidants (such as O_3), and catalytic, metal-containing particulates. A typical residence time for atmospheric SO_2 is approximately 4 days; this allows much emitted SO_2 to be transported substantial distances from points of emission before oxidation, deposition or exposure to ground-level plants occurs (this is known as the long-range transport of air pollutants).

The largest anthropogenic emissions of SO_2 result from the combustion of sulphur-containing fossil fuels (primarily coal and residual fuel oils), which account for over 50% of the total emissions. The remaining emissions of SO_2 are mostly associated with various manufacturing processes and the smelting of sulphide ores. Since the turn of the 20th century, global anthropogenic emissions of SO_2 have increased more than fivefold (Freedman, 1995). As a result of changes in technology and increased industrialization and urbanization during this period, a much larger portion of the landscape is now being affected by SO_2 pollution than before. However, abatement policies implemented in Western Europe and North America since the early 1960s, and more recently in Japan, have decreased anthropogenic sulphur emissions in those regions (see Bull and Fenech, this volume). The installation of tall emissions stacks at major point-sources has also reduced many of the more severe local effects by promoting better dispersion before acute ground-level exposures occur; this has been referred to as the 'dilution solution to air pollution'. The tall stacks have resulted in SO_2 pollution becoming less acute, but more broadly regionalized. The use of flue-gas desulphurization technologies is more expensive than building tall stacks, but the former helps to reduce the effects of long-range air pollution transport. Conversely, SO_2 emissions have steadily risen in Russia and parts of Asia, particularly in China, due to increasing levels of coal combustion. However, in the early 1990s, emission controls began to be introduced in some of the formerly communist countries of Eastern Europe (Berresheim *et al.*, 1995), and emissions at some of the Russian point-sources have declined following reductions in industrial outputs (see Rigina and Kozlov, this volume).

Concentrations of SO_2 in clean air are generally less than 2 ppb. Annual means of SO_2 in urban air are typically < 15–20 ppb, while sites near major point-sources of emissions can reach values of 1 ppm and much higher during periods of intense fumigation (known as pollution episodes). SO_2 can be phytotoxic to sensitive plants at ambient concentrations of > 50 ppb (Fowler, 1992). Areas which have been subjected to high levels of SO_2 pollution near point-sources, such as those near the metal smelters at Sudbury in Canada, have been ecologically devastated (Freedman, 1995; Gunn, 1995) and are only now beginning to show signs of recovery (see Winterhalder, this volume).

SO_2 can cause acute and chronic injury to vegetation. Short-term exposures may result in acute tissue damage such as necrosis and foliage death. Longer-term exposures may result in chronic symptoms such as chlorosis (yellowing of foliage), premature abscission of foliage and reductions in production. Loss of yield may also occur in the absence of other apparent signs of injury (this is known as 'hidden injury'). Phytotoxic thresholds of SO_2 vary depending upon the susceptibility of species and local genotypes, as well as interacting environmental factors such as season, drought, nutrient supply and the presence of other pollutants. Coniferous tree species tend to be more sensitive than deciduous species (Singh *et al.*, 1991), which is possibly related to the higher doses received by coniferous species during winter, but further generalizations about the relative sensitivities of particular species are very difficult to make.

Short-term, periodic exposures typical of fumigations occurring in the vicinity of large point-sources can cause acute injury to most plant species at a concentration of 0.7 ppm SO_2 over a 1-h period, or at 0.18 ppm SO_2 over an 8-h period (Shriner, 1990). Most tree species will be affected by fumigations with an average concentration of > 0.17 ppm occurring 10% of the time during their growing season (Roberts, 1984). This generalized threshold is, however, lower for more sensitive species or varieties. Regional effects have also been linked to the deposition of SO_2 over large areas (i.e. > 10^4 km^2); however, the causal relationships are much less clearly distinguished (e.g. Schulze, 1989; Fowler, 1992). The laboratory and field studies reviewed by

Roberts (1984) indicated the following levels of chronic exposure to SO_2 at which yield is likely to be affected in plant species:

1. 76–150 ppb for 1–3 months results in significant decreases in yield for most species.
2. 38–76 ppb for several months results in small decreases in yield for some sensitive species.
3. < 38 ppb results in minor decreases in yield for some hypersensitive species, while others experience no damage or may benefit from sulphur fertilization.

Changes in community structure may occur in chronically exposed areas due to selection for species of plants which are more tolerant of SO_2 (Smith, 1990).

As is the case with all pollutants from large point-sources, a directional gradient of decreasing forest damage is commonly observed along radial transects from large point-sources of SO_2 emissions. This pattern is related to the progressive dispersal and dilution of pollutants as influenced by prevailing winds and topography. For example, reduced growth of Scots pine was observed up to 30 km south-west of a nickel–copper smelter on the Kola Peninsula in Russia, with the effects being most severe closer to the point-source (Nöjd and Reams, 1996; Rigina and Kozlov, this volume). This pattern of damage is especially well documented near Sudbury, Canada (Struik, 1974; Freedman and Hutchinson, 1980; Winterhalder, 1995). In both cases, high levels of sulphur emissions were the primary cause of damage to forest vegetation, although heavy metals also probably contributed.

Particulate elemental sulphur (S) can be an important pollutant in the vicinity of sour-gas processing plants (i.e. methane rich in H_2S), as a result of fugitive releases (Mayo *et al.*, 1992). This industry may also emit SO_2 through the incineration of H_2S unrecovered as elemental sulphur. It is thought that many of the effects on surrounding vegetation are related to changes in soil chemistry following the metabolism of elemental sulphur by species of the naturally occurring soil bacterium *Thiobacillus*. S is oxidized to sulphuric acid by these bacteria, causing extreme soil acidification (Legge *et al.*, 1988). Dramatic declines in the abundance and diversity of higher plants, bryophytes and lichens have been documented, for example, in Alberta, Canada (Kennedy *et al.*, 1988). However, it is difficult to separate the effects of S from those of co-occurring pollutants.

2.2 Nitrogen gases

The most significant atmospheric nitrogen compounds affecting forests are the NO_x gases (consisting of nitric oxide (NO) and nitrogen dioxide (NO_2)) and ammonia (NH_3). These gases all have natural sources of emissions, but they may create local or regional ecological problems when emitted in large amounts as a result of human activities. Anthropogenic formation of NO_x gases occurs during the combustion of fossil fuels, through the oxidation of organic-N and atmospheric N_2, the latter occurring especially in the internal combustion engines of automobiles. It is primarily NO which is produced through oxidation during combustion processes, while NO_2 forms secondarily in the atmosphere by the photo-oxidation of emitted NO. NH_3 is emitted as fugitive emissions from livestock feedlots, fur farms and solid-waste disposal sites, and also during the combustion of fossil fuels. NH_3 is oxidized to NO_x in the atmosphere. In addition, NO, NO_2 and NH_3 can be dry-deposited to terrestrial and aquatic surfaces, and contribute to acidification when they become oxidized to nitrate (NO_3^-).

NO_x gases rarely reach concentrations sufficient to cause acute foliar injury, except in extremely polluted conditions close to industrial point-sources. Acute injury to plants by NO_2 can occur at exposures of 20 ppm for 1 h, or at 2 ppm for a 48-h period (Shriner, 1990). Some of the effects of exposures to elevated levels of these gases include the potential for decreased frost hardiness, increased insect and fungal attraction to the trees, nutrient imbalances, and increased susceptibility to wind stress due to shallow root growth if nitrogen accumulates in the soil surface (Treshow, 1984; Smith, 1990).

In general, trees are tolerant of NO_x gases, and growth may be directly stimulated by increased deposition. NO_x pose their most significant risk to forests through their roles in the synthesis of photochemical air pollutants, the acidification of precipitation, the acidification of ecosystems by wet and dry deposition of NO_x and NH_3, and by the nitrogen saturation of ecosystems by high rates of deposition of these gases.

2.3   Photochemical air pollutants

Ozone (O_3) is by far the most damaging of the photochemical air pollutants, followed by peroxyacetyl nitrate (or PAN). O_3 is considered to be the air pollutant of greatest concern in North America because of its regional effects on forest trees and other kinds of terrestrial vegetation, including agroecosystems (Lefohn et al., 1997). Tropospheric ozone is a secondary pollutant, which is formed through the photo-oxidation of VOCs (volatile organic compounds, including hydrocarbons) in the presence of NO_x. NO and hydrocarbons are present naturally, but they are also emitted in large quantities by motor vehicles and industries as by-products of combustion processes. In the presence of high levels of solar radiation, NO is oxidized to NO_2 which, through further photochemical reactions, produces free O atoms, which react with O_2 to create O_3.

Atmospheric concentrations of O_3 in areas affected by photochemical pollution tend to vary predictably at different times of the day, depending upon the rates of emission of NO_x and hydrocarbons, the intensity of sunlight and the occurrence of stable atmospheric inversions (Singh et al., 1978; Sanz and Millán, this volume). Ozone levels also vary seasonally and annually, primarily according to meteorological conditions. For example, during hot, sunny summers O_3 concentrations are typically higher than during cool, rainy summers in the same area (Miller et al., 1994; De Leeuw and Van Zantvoort, 1997).

Ozone is known to cause a range of damage to forest vegetation. In highly polluted areas, acute injury occurs typically when plants are exposed to O_3 at concentrations of 200–300 ppb for a period of 2–4 h (Shriner, 1990). Such exposures typically result in foliar damage, which reduces the area of leaf tissue capable of photosynthesis and thus decreases the rate of productivity, while also increasing dark respiration (which causes additional decreases in net productivity) (Kim and Kim, 1997). Sensitive species of plants, including some conifers, may experience acute or hidden injuries at concentrations as low as 80 ppb over a 12-h exposure. Controlled-exposure studies have shown that average daytime O_3 concentrations of ≥ 100 ppb can alter carbon allocation processes and nutrient status, reduce chlorophyll levels in foliage and accelerate foliar senescence (Shriner, 1990). Trees stressed by oxidant injury may also be more vulnerable to insect attack (Chappelka et al., 1997).

In the valley cradling Mexico City, two sensitive species of pine (*Pinus hartwegii* and *P. leiphylla*) exhibit damage diagnostic of ozone injury. These include chlorotic banding and mottling of the foliage, premature needle loss and branch mortality (De Bauer and Krupa, 1990). The observed effects decrease along a spatial gradient from the outskirts of Mexico City to a distance of 30 km. Similar observations have been made in southern California (see Miller and Arbaugh, this volume).

Ozone can be transported over rather long distances from urban sources to rural, forested areas, where extended periods of exposure at relatively moderate concentrations may affect vegetation. Effects of this sort have been well documented, for example, in the San Bernardino Mountains east of the Los Angeles basin in southern California (Miller et al., 1989; Barnard and Lucier, 1991; Miller and Arbaugh, this volume). Effects on trees may include acute leaf injury, reductions in productivity and yield, and changes in the tolerance of plants to other biotic and abiotic stresses (Bartholomay et al., 1997; Fuhrer et al., 1997). Over several growing seasons, incremental growth declines have been shown to occur in seedlings of ponderosa pine (*Pinus ponderosa* Dougl.) at 24-h average exposures to only 60 ppb (Takemoto et al., 1997). These effects may lead to changes in seedling recruitment and species composition of stands (Constable and Taylor, 1997) and may alter the genetic composition of tree populations (Taylor, 1994). Less is understood regarding the longer-term responses of mature trees to ozone exposures; almost all controlled studies have been carried out on seedlings and saplings.

Climatic factors have an effect on the response of forest vegetation to ozone exposure. McLaughlin and Downing (1995) found that mature loblolly pines

(*Pinus taeda* L.) experienced reduced rates of stem growth when exposed to average daily O_3 of ≥ 40 ppb. This effect was most evident when the O_3 exposure was associated with low soil moisture and high air temperature.

Preliminary attempts have been made to determine the relative sensitivity of various tree species to O_3 in the eastern USA (Hogsett *et al.*, 1997). Aspen (*Populus* spp.) and black cherry (*Prunus serotina* Ehrh.) appear to be the most sensitive. Tulip poplar (*Liriodendron tulipifera* L.), loblolly pine, eastern white pine (*Pinus strobus* L.) and sugar maple (*Acer saccharum* Marsh.) are moderately sensitive. Virginia pine (*Pinus virginiana* Mill.) and red maple (*Acer rubrum* L.) are the least sensitive of the species considered. However, such categorizations cannot necessarily be applied to other areas, as local populations may vary in sensitivity. Furthermore, the species that are the most tolerant of exposure to ozone may not be the most tolerant of other air pollutants.

2.4   Fluorides

The largest portion of anthropogenic emissions of fluoride gases consist of hydrogen fluoride (HF) and silicon tetrafluoride (SiF_4). These originate predominantly from coal combustion, phosphate fertilizer processing, aluminium reduction, and brick and tile manufacturing. Effects on forest vegetation mostly occur close to large point-sources (Klumpp, 1996). However, HF is relatively easily removed from flue gases, and can therefore be controlled if investments are made in the appropriate emissions-reduction technology (Halbwachs, 1984).

Atmospheric concentrations of fluoride in remote areas are typically less than 0.3 ppb, while urban areas may measure on the order of < 3 ppb (Urone, 1986). Concentrations in air close to large industrial sources can be much higher. For example, an area in Florida close to an operation mining and processing phosphate rock was found to contain up to 250 ppb in the surrounding air (National Academy of Sciences, 1971; Freedman, 1995).

Fluoride accumulates in the tissues of plants, which exhibit a wide range of sensitivity to the substance. Sensitive plants can suffer acute injury if exposed to over 4–5 ppb for a 1-day period, while more tolerant species require exposure to > 13 ppb. Over a 1-month period, exposure to concentrations of 0.6 ppb and 1–4 ppb can cause acute injury to sensitive and tolerant species respectively. At foliar concentrations of 20–150 ppm F, sensitive species will exhibit symptoms of acute injury. Others are capable of tolerating an accumulation of > 4000 ppm in their tissue (Urone, 1986; Freedman, 1995).

In the vicinity of a phosphorus plant in Newfoundland, Canada, conifer species such as balsam fir (*Abies balsamea* (L.) Mill.), black spruce (*Picea mariana* (Mill.) BSP) and tamarack (*Larix laricina* (Du Roi) K. Koch) were found to be more susceptible to damage from fluoride pollution than certain tolerant hardwood species, e.g. white birch (*Betula papyrifera* Marsh.) and green alder (*Alnus crispa* (Ait.) Pursh.) (Sidhu and Staniforth, 1986). The effects observed in conifers when foliar accumulations exceeded 20 ppm included various degrees of foliar damage (chlorosis, necrosis, needle damage, defoliation), reductions in seed size, percentage germination, numbers of seeds per cone, number of cones per tree, number of fertile trees, and distortion and mortality of cones.

2.5 Metals

The trace metals that occur naturally in rocks and soils are mostly in forms which are not readily bioavailable. In contrast, metals in the form of relatively soluble oxides and halides are released to the environment through the combustion of fossil fuels, industrial processes and waste disposal activities such as incineration (Table 2.3). All heavy metals are potentially phytotoxic in high doses. The characteristics of the receiving soil (especially pH, oxidation–reduction potential, aeration, temperature, and concentrations of inorganic anions and cations, particulate and soluble organic matter, clay minerals and salinity) will influence the availability and thus toxicity of most metals (Maxwell, 1995).

Metals are deposited from the atmosphere primarily as particulates in dry deposition, wet deposition (carried by precipitation), and in fog and mists (occult deposition) (Ross, 1994). A few metals, notably mercury, are also dry-deposited as vapours. Long-range transport in the atmosphere, particularly of ionic species, can occur. However, local and regional metal pollution co-occurring with SO_2 around metal smelters and other large industrial facilities, with serious effects on forests, has been much more clearly documented. For example, an area of about 500 km^2 has been polluted by metals emitted by smelters in the Sudbury area (Gunn, 1995). An area totalling 39,000 km^2 surrounding the industrial centres of Monchegorsk and Nikel on the Kola Peninsula has been polluted by a combination of nickel, copper and SO_2 emitted in large part by metal smelters (Tikkanen and Niemelä, 1995; Evdokimova, this volume; Rigina and Kozlov, this volume). In another case, damage to vegetation surrounding a brass works in Gusum, Sweden, can be attributed

Table 2.3. Sources of metals introduced to terrestrial vegetation through atmospheric deposition. (After Ross, 1994.)

Source	Metals
Pyrometallurgical industries	As, Cd, Cr, Cu, Mn, Ni, Pb, Sb, Tl, Zn
Urban industrial, especially incineration	Cd, Cu, Pb, Sn, Hg, V
Motor vehicles	Mo, Pb (with Br, Cl), V
Other fossil-fuel combustion, including power plants	As, Pb, Sb, Se, U, V, Zn, Cd

primarily to extremely high concentrations of zinc and copper accumulated in surface organic matter during a period of more than three centuries of operation, SO_2 not being an important factor at that site (Tyler, 1984).

Heavy metal contamination of soils can cause severe stress in plants, particularly through exposures in the root zone. Direct, acute toxicity of heavy metals is known to occur in forest vegetation predominantly in the vicinity of large point-sources, particularly smelters and metal refineries. Trees and many herbaceous species may be virtually absent in soils heavily contaminated with metals from industrial sources. Copper, nickel and zinc are most frequently associated with direct phytotoxicity (Ross, 1994). Cadmium, cobalt and lead have also been shown to cause phytotoxic effects, although this is less common. Symptoms of metal toxicity may include stunting of growth and chlorosis. Restriction of lateral growth from root meristems is often the first sign of damage. The inhibition of root growth may impede the uptake of phosphorus, potassium and iron, thereby restricting normal metabolism (Smith, 1990).

In many herbaceous species, the evolution of metal-resistant ecotypes is known to occur rapidly (Turner, 1994). Species of birch (*Betula pendula* Roth. and *B. pubescens* Ehrh.) have been shown to evolve populations tolerant to high levels of zinc present in mine wastes, by resisting the uptake of the metal (Turner, 1994; Freedman, 1995). Watmough and Hutchinson (1997) suggested that individual red maples may have the ability to develop an adaptive tolerance, within their lifetimes, to high levels of metals in soil, particularly nickel. Evidence for this was the development of resistant cell lines within the shoot meristems.

2.6   Deposition of acidifying substances from the atmosphere

The acidification of forest ecosystems (i.e. increasing concentrations of hydrogen ions (H^+) in soil and water) can result from acidic precipitation, and from the deposition and subsequent oxidation of substances such as SO_2, NO_x, NH_3 and NH_4^+. These may reach vegetative surfaces as wet deposition in the form of rain, fog or snow. Atmospheric gases and particulates may also dry-deposit during periods between precipitation events. Dry deposition typically accounts for the larger proportion of the total input of acidifying substances in heavily industrialized and urbanized areas, except in very moist climates where wet deposition may predominate.

Intensified acidification above normal, background levels, caused by large inputs of anthropogenic pollutants, may be associated indirectly with forest decline (Haines and Carlson, 1989). Acidification may be caused when hydrogen ions increase in concentration to electrochemically balance wet-deposited sulphate (SO_4^{2-}) and that produced through the oxidation of dry-deposited

SO_2, and the NO_3^- produced by wet deposition and the oxidation of NO_x, NH_3 and NH_4^+ (Freedman, 1995).

Large inputs of acidifying substances to weakly buffered soils can lead to increased cation leaching and the generation of potentially toxic concentrations of the aluminium ions Al^{3+} and $Al(OH)^{2+}$. This can cause damage to root systems, characterized by cellular disruption associated with inhibited calcium and magnesium uptake (Fowler, 1992). Poor root development results in additional nutrient and water stress in trees. Acidification also exacerbates the effects of heavy metal pollution, by increasing the solubility and thereby the bioavailability of metals present in soil.

3   Predicting the risks of air pollution to forests

Attempts have been made to establish critical exposure levels for some important air pollutants, above which damage to vegetation (such as reductions in productivity) can be expected to occur (for discussion see, e.g. OECD, 1990; Fuhrer and Achermann, 1994; Bull, 1995; Sanders *et al.*, 1995; Fuhrer *et al.*, 1997). The use of values established for agricultural crops has been particularly important in developing pollution abatement strategies within and between nations (Bull, 1995). However, the concept of critical levels is difficult to apply to mature forest trees, since there are limited available observational and experimental data upon which to base them. The most reliable data that exist for the effects of various exposures to air pollutants are derived from studies on seedlings or saplings, or they are specific to particular regions, e.g. north-western Europe. Until they are better refined, the universal application of general critical levels for risk assessment is problematic, as they are not yet able to reflect the variations in sensitivity inherent in plant species, populations and communities in various biogeographic regions (Fuhrer *et al.*, 1997). The specific effects of land management practices, soils, moisture availability and climate in various regions are also important variables that are not readily integrated into simple criteria for critical exposures.

More extensive field data are needed to better understand the potential effects of air pollution on forest vegetation. Methods have been developed to conduct risk assessments through modelling of known effects on plants to known ambient exposures (e.g. Kim and Kim, 1997; Constable and Taylor, 1997; Hogsett *et al.*, 1997). The use of sensitive bioindicators to monitor air pollution is another useful approach which can limit the amount of data needed to determine thresholds of ecological responses to pollution (National Research Council, 1989; Ali, 1993). Soil biological variables such as respiration rate and microbial biomass have also been investigated for use as early indicators of the ecological effects of air pollution (Vanhala *et al.*, 1996; Evdokimova, this volume).

In summary, some important general observations of the effects of air pollution on forests in heavily industrialized and urbanized areas are as follows:

1. Direct damage to vegetation near large point-sources is characterized by a spatial pattern of exponentially decreasing intensity with increasing radial distance.

2. Acute effects on vegetation may occur in areas immediately surrounding large sources of emitted pollutants, while lower-level, regional effects may be much more widespread, but difficult to document.

3. The direct and indirect effects of air pollutants in forest ecosystems are dependent upon numerous complex variables, including the intensity of exposure, mitigating environmental conditions and the inherent susceptibility of affected populations, species and communities.

References

Ali, E.A. (1993) Damage to plants due to industrial pollution and their use as bioindicators in Egypt. *Environmental Pollution* 81, 251–255.

Barnard, J.E. and Lucier, A.A. (1991) Changes in forest health and productivity in the United States and Canada. In: Irving, P.M. (ed.) *Acidic Deposition: State of Science and Technology. Summary Report of the U.S. National Acid Precipitation Assessment Program.* National Acid Precipitation Assessment Program, Washington, DC, pp. 135–138.

Bartholomay, G.A., Eckert, R.T. and Smith, K.T. (1997) Reductions in tree-ring widths of white pine following ozone exposure at Acadia National Park, Maine, U.S.A. *Canadian Journal of Forest Research* 27, 361–368.

Berresheim, H., Wine, P.H. and Davis, D.D. (1995) Sulfur in the atmosphere. In: Singh, H.B. (ed.) *Composition, Chemistry and Climate of the Atmosphere.* Van Nostrand Reinhold, New York, pp. 251–307.

Bormann, F.H. (1982) The effects of air pollution on the New England landscape. *Ambio* 11, 338–346.

Bull, K.R. (1995) Critical loads: possibilities and constraints. *Water, Air, and Soil Pollution* 85, 201–212.

Bussotti, F., Cenni, E., Cozzi, A. and Ferretti, M. (1997) The impact of geothermal power plants on forest vegetation: a case study at Travale (Tuscany, Central Italy). *Environmental Monitoring and Assessment* 45, 181–194.

Chappelka, A., Renfro, J., Somers, G. and Nash, B. (1997) Evaluation of ozone injury on foliage of black cherry (*Prunus serotina*) and tall milkweed (*Asclepias exaltata*) in Great Smoky Mountains National Park. *Environmental Pollution* 95, 13–18.

Constable, J.V.H. and Taylor, G.E., Jr (1997) Modeling the effects of elevated tropospheric O_3 on two varieties of *Pinus ponderosa. Canadian Journal of Forest Research* 27, 527–537.

Darrall, N.M. (1989) The effects of air pollutants on physiological processes in plants. *Plant, Cell and Environment* 12, 1–30.

De Bauer, M. and Krupa, S.V. (1990) The Valley of Mexico: summary of observational studies on its air quality and effects on vegetation. *Environmental Pollution* 65, 109–118.

De Leeuw, F.A.A.M. and Van Zantvoort, E.D.G. (1997) Mapping of exceedances of ozone critical levels for crops and forest trees in the Netherlands: preliminary results. *Environmental Pollution* 96, 89–98.

Fowler, D. (1992) Effects of acidic pollutants on terrestrial ecosystems. In: Radojević, M. and Harrison, R.M. (eds) *Atmospheric Acidity: Sources, Consequences and Abatement*. Elsevier Applied Science, London, pp. 341–362.

Freedman, B. (1995) *Environmental Ecology. The Ecological Effects of Pollution, Disturbance, and Other Stresses*, 2nd edn. Academic Press, San Diego.

Freedman, B. and Hutchinson, T.C. (1980) Long-term effects of smelter pollution at Sudbury, Ontario on forest community composition. *Canadian Journal of Botany* 58, 2123–2140.

Fuhrer, J. and Achermann, B. (eds) (1994) *Critical Levels for Ozone – a UN-ECE Workshop Report*. Schriftenreihe der FAC Liebefeld 16. Swiss Federal Research Station for Agricultural Chemistry and Environmental Hygiene, Liebefeld-Bern.

Fuhrer, J., Skärby, L. and Ashmore, M.R. (1997) Critical levels for ozone: effects on vegetation in Europe. *Environmental Pollution* 97, 91–106.

Gunn, J.M. (ed.) (1995) *Restoration and Recovery of an Industrial Region: Progress in Restoring the Smelter-Damaged Landscape near Sudbury, Canada*. Springer-Verlag, New York.

Haines, B.L. and Carlson, C.L. (1989) Effects of acidic precipitation on trees. In: Adriano, D.C. and Johnson, A.H. (eds) *Acidic Precipitation*, Vol. 2, *Biological and Ecological Effects*. Springer-Verlag, New York, pp. 1–27.

Halbwachs, G. (1984) Organismal responses of higher plants to atmospheric pollutants: Sulphur dioxide and fluoride. In: Treshow, M. (ed.) *Air Pollution and Plant Life*. John Wiley & Sons, Toronto, pp. 175–214.

Hansen, J.R., Hansson, R. and Norris, S. (1996) *The State of the European Arctic Environment*. EEA Environmental Monograph No. 3. European Environment Agency, Copenhagen.

Heck, W.W. and Brandt, C.S. (1977) Effects on vegetation: native, crops, forests. In: Stern, A.C. (ed.) *Air Pollution*, Vol. II, *The Effects of Air Pollution*, 3rd edn. Academic Press, New York, pp. 157–229.

Hogsett, W.E., Weber, J.E., Tingey, D., Herstrom, A., Lee, E.H. and Laurence, J.A. (1997) Environmental auditing: an approach for characterizing tropospheric ozone risk to forests. *Journal of Environmental Management* 21, 105–120.

Kato, N. and Akimoto, H. (1992) Anthropogenic emissions of SO_2 and NO_x in Asia: emission inventories. *Atmospheric Environment* 26A, 2997–3017.

Kennedy, K.A., Addison, P.A. and Maynard, D.G. (1988) Effect of elemental sulphur on the vegetation of a lodgepole pine stand. *Environmental Pollution* 51, 121–130.

Kim, J.W. and Kim, J.-H. (1997) Modelling the net photosynthetic rate of *Quercus mongolica* stands affected by ambient ozone. *Ecological Modelling* 97, 167–177.

Klumpp, A. (1996) Assessment of the vegetation risk by fluoride emissions from fertilizer industries at Cubatao, Brazil. *Science of the Total Environment* 96, 219–229.

Lefohn, A.S., Jackson, W., Shadwick, D.S. and Knudsen, H.P. (1997) Effect of surface ozone exposures on vegetation growth in the Southern Appalachian

Mountains: identification of possible areas of concern. *Atmospheric Environment* 31, 1695–1708.

Legge, A.H. and Krupa, S.V. (eds) (1986) *Air Pollutants and their Effects on the Terrestrial Ecosystem*. John Wiley & Sons, New York.

Legge, A.H., Bogner, J.C. and Krupa, S.V. (1988) Foliar sulphur species in pine: a new indicator of a forest ecosystem under pollution stress. *Environmental Pollution* 55, 15–27.

Mayo, J.M., Legge, A.H., Yeung, E.C., Krupa, S.V. and Bogner, J.C. (1992) The effects of sulphur gas and elemental sulphur dust deposition on *Pinus contorta × Pinus banksiana*: cell walls and water relations. *Environmental Pollution* 76, 43–50.

Maxwell, C.D. (1995) Acidification and metal contamination: implications for the soil biota of Sudbury. In: Gunn, J.M. (ed.) *Restoration and Recovery of an Industrial Region: Progress in Restoring the Smelter-Damaged Landscape near Sudbury, Canada*. Springer-Verlag, New York, pp. 219–231.

McLaughlin, S.B. and Downing, D.J. (1995) Interactive effects of ambient ozone and climate measured on growth of mature forest trees. *Nature* 374, 252–254.

Miller, P.R. and McBride, J.R. (1975) Effects of air pollutants on forests. In: Mudd, J.B. and Kozlowski, T.T. (eds) *Responses of Plants to Air Pollution*. Academic Press, New York, pp. 195–235.

Miller, P.R., McBride, S.R., Schilling, S.L. and Gomez, A.P. (1989) Trend of ozone damage to conifer forests between 1974 and 1988 in the San Bernardino Mountains of southern California. In: Olson, R.K. and Lefohn, A.S. (eds) *Effect of Air Pollution on Western Forests*. Transaction Series. Air and Waste Management Association, Pittsburgh, pp. 309–324.

Miller, P.R., De Bauer, M.L., Quevedo Nolasco, A. and Hernandez Tejeda, T. (1994) Comparison of ozone exposure characteristics in forested regions near Mexico City and Los Angeles. *Atmospheric Environment* 28, 141–148.

National Academy of Sciences (1971) *Fluorides*. Committee on Biologic Effects of Atmospheric Pollutants. National Research Council, Division of Medical Sciences, Washington, DC.

National Research Council (US) (1989) *Biologic Markers of Air-pollution Stress and Damage in Forests*. National Academy Press, Washington, DC.

Nöjd, P. and Reams, G.A. (1996) Growth variation of Scots pine across a pollution gradient on the Kola Peninsula, Russia. *Environmental Pollution* 93, 313–325.

Nriagu, J.O. (ed.) (1984) *Environmental Impacts of Smelters*. John Wiley & Sons, New York.

OECD (1990) *Control Strategies for Photochemical Oxidants across Europe*. Organisation for Economic Co-operation and Development, Paris.

Peterson, D.J. (1993) *Troubled Lands: the Legacy of the Soviet Environmental Destruction*. Westview Press, Boulder, Colorado.

Pignata, M.L., Cañas, M.S., Carreras, H.A. and Orellana, L. (1997) Exploring chemical variables in *Ligustrum lucidum* Ait. f. *tricolor* (Rehd.) Rehd. in relation to air pollutants and environmental conditions. *Journal of Environmental Management* 21, 793–801.

Pryde, P.R. (1991) *Environmental Management in the Soviet Union*. Cambridge University Press, Cambridge.

Roberts, T.M. (1984) Effects of air pollutants in agriculture and forestry. *Atmospheric Environment* 18, 629–652.

Ross, S.M. (ed.) (1994) *Toxic Metals in Soil–Plant Systems.* John Wiley & Sons, Chichester.

Sanders, G.E., Skärby, L., Ashmore, M.R. and Fuhrer, J. (1995) Establishing critical levels for the effects of air pollution on vegetation. *Water, Air, and Soil Pollution* 85, 189–200.

Schulze, E.D. (1989) Air pollution and forest decline in a spruce (*Picea abies*) forest. *Science* 244, 776–783.

Seinfeld, J.H. and Pandis, S.N. (1998) *Atmospheric Chemistry and Physics: from Air Pollution to Climate Change.* John Wiley & Sons, New York.

Shriner, D.S. (1990) Responses of vegetation to atmospheric deposition and air pollution. In: Anon. (ed.) *Acidic Deposition: State of Science and Technology.* Vol. III. *Terrestrial, Materials, Health and Visibility Effects.* US Government Printing Office, Washington, DC, pp. 18-1–18-206.

Sidhu, S.S. and Staniforth, R.J. (1986) Effects of atmospheric fluorides on foliage, and cone and seed production in balsam fir, black spruce, and larch. *Canadian Journal of Botany* 64, 923–931.

Singh, H.B., Ludwig, F.L. and Johnson, W.B. (1978) Tropospheric ozone: concentrations and variabilities in clean remote atmospheres. *Atmospheric Environment* 12, 2185–2196.

Singh, S.K., Rao, D.N., Agrawal, M., Pandey, J. and Narayan, D. (1991) Air pollution tolerance index of plants. *Journal of Environmental Management* 32, 45–55.

Smith, W.H. (1990) *Air Pollution and Forests: Interaction Between Air Contaminants and Forest Ecosystems.* 2nd Edn. Springer-Verlag, New York.

Strand, M. (1993) Photosynthetic activity of Scots pine (*Pinus sylvestris* L.) needles during winter is affected by exposure to SO_2 and NO_2 during summer. *New Phytologist* 123, 133–141.

Struik, H. (1974) *Photo Interpretive Study to Assess and Evaluate Vegetational Changes in the Sudbury Area.* Internal Report, Department of Lands and Forests, Ottawa, Ontario.

Takemoto, B.K., Bytnerowicz, A., Dawson, P.J., Morrison, C.L. and Temple, P.J. (1997) Effects of ozone on *Pinus ponderosa* seedlings: comparison of responses in the first and second growing seasons of exposure. *Canadian Journal of Forest Research* 27, 23–30.

Taylor, G.E. (1994) Role of genotype in the response of loblolly pine to tropospheric ozone: effects at whole-tree, stand, and regional level. *Journal of Environmental Quality* 23, 63–82.

Tikkanen, E. and Niemelä, I. (eds) (1995) *Kola Peninsula Pollutants and Forest Ecosystems in Lapland. Final Report of The Lapland Forest Damage Project.* Finland's Ministry of Agriculture and Forestry and The Finnish Forest Research Institute, Rovaniemi, Finland.

Treshow, M. (ed.) (1984) *Air Pollution and Plant Life.* John Wiley & Sons, Chichester.

Turner, A.P. (1994) The responses of plants to heavy metals. In: Ross, S.M. (ed.) *Toxic Metals in Soil-Plant Systems.* John Wiley & Sons, Chichester, pp. 153–187.

Tyler, G. (1984) The impact of heavy metal pollution on forests: a case study of Gusum, Sweden. *Ambio* 13, 18–24.

Urone, P. (1986) The pollutants. In: Stern, A.C. (ed.) *Air Pollution,* 3rd edn, Vol. 4. Academic Press, Orlando, Florida, pp. 1–60.

Vanhala, P., Kiikkila, O. and Fritze, H. (1996) Microbial responses of forest soil to moderate anthropogenic air pollution: a large scale field survey. *Water, Air, and Soil Pollution* 86, 173–186.

Watmough, S.A. and Hutchinson, T.C. (1997) Metal resistance in red maple (*Acer rubrum*) callus cultures from mine and smelter sites in Canada. *Canadian Journal of Forest Research* 27, 693–700.

Winterhalder, K. (1995) Early history of human activities in the Sudbury area and ecological damage to the landscape. In: Gunn, J.M. (ed.) *Restoration and Recovery of an Industrial Region: Progress in Restoring the Smelter-Damaged Landscape near Sudbury, Canada.* Springer-Verlag, New York, pp. 17–31.

Wolfenden, J. and Mansfield, T.A. (1991) Physiological disturbances in plants caused by air pollutants. *Proceedings of the Royal Society of Edinburgh* 97B, 117–138.

General Characteristics of the Damage to Forest Ecosystems in Russia Caused by Industrial Pollution

3

N.P. Vassilieva,[1] M.L. Gytarsky[2] and R.T. Karaban[2]

[1]All-Russian Institute for Nature Conservation, Moscow, Russian Federation; [2]Institute of Global Climate and Ecology, Moscow, Russian Federation

Industrial emissions cause local damage to forests in the vicinity of large enterprises in various parts of the country. The area of possible forest decline in Russia attributable to air pollution is probably about 1.3 million hectares. Non-ferrous metallurgy and power generation are the major sources of phytotoxic atmospheric emissions. The main method of forest vitality assessment that is applied over highly polluted regions of the country consists of monitoring, although a number of different forest monitoring systems exist. No significant improvement in forest vitality in damaged areas is expected in the near future.

1   Introduction

Forest ecosystems are sensitive to environmental quality. Increased levels of industrial pollution have caused damage to forests over local regions of the Russian Federation, and continue to do so. Adverse effects endanger ecosystem integrity, cause the dieback of individual species, and finally result in the decline of stands in areas subject to very high concentrations of contaminants. Consequently, continuous observations of forest health are especially important in the heavily polluted regions of the country. Numerous criteria and methodologies are currently used by different institutions in Russia for various types of forest health assessments. This work provides an overview of the problems in Russia and examines the existing types of forest monitoring based on the common parameters and methods applied by most of the investigators.

Subsequent chapters deal with case studies, specifically the smelters of the Kola Peninsula and Noril'sk (Chapters 4, 5 and 6 of this volume).

2   Monitoring systems and methods

Monitoring is the main method currently applied for forest health assessment in the heavily polluted regions of Russia. A national system of environmental monitoring has been developing since the 1970s. In 1993, the Russian Government decided to organize a Unified System for State Ecological Monitoring (USSEM) over the territory of the Russian Federation. It is still under development. The USSEM will consolidate almost 20 different ministries and agencies, all of which perform continuous observations and monitoring of natural phenomena and human activities relevant to the environment as a whole. The general coordination of activities within the framework of the Unified System of State Ecological Monitoring is undertaken by the State Committee for Environmental Conservation of the Russian Federation (Anon., 1994, 1997a).

Forest monitoring is an important part of the USSEM. Table 3.1 presents examples of the institutions, ministries and agencies involved in various types of forest monitoring in the Russian Federation. Official responsibility for forest control lies with the Federal Forestry Service of Russia. Regular observations provide valuable information on the current state of forests and their responses to the adverse impacts of industrial pollution. Generally, observations are made by the personnel of the responsible institutions, using their own networks, although specialists from a number of different institutions could be engaged in the joint implementation of monitoring activities. The common use of networks would also be possible in the case of bilateral, multilateral or other types of cooperation between organizations. Such joint efforts provide good opportunities for an interdisciplinary and thorough approach to observations, as well as ensuring the adequate qualifications of the personnel involved. The results that have been obtained to date are summarized annually in the State Reports on the Environmental Situation in the Russian Federation and other analytical documents (Anon., 1994, 1997a, b).

The visual assessment of forest condition and the selective measurement of a number of tree parameters in temporary and permanent sample plots are the most commonly used methods of forest monitoring and control. Visual assessments include the determination of the health of trees based on a special scale of tree health and the identification of forest decline based on signs of defoliation, chlorosis (discoloration), necrosis, insect attacks and phytopathogen infestations. The obligatory parameters are the diameter, height and radial growth of trees. Measurements for pine trees (*Pinus* spp.), which are known to be highly sensitive to air pollution, include leader and upper lateral shoot lengths and the age of needles on top and lateral shoots (Mozolevskaya *et al.*, 1984; Anon., 1987, 1992).

Sampling is an important part of the monitoring process. Samples of the foliage of dominant trees, ground vegetation and soil are collected, together with wet and dry deposited contaminants. The samples are analysed for the dominant chemical elements in the industrial emissions of local enterprises. All observations, and the collection, treatment and analysis of samples, are performed according to inter-institutional and modified conventional methodological manuals and guidelines (Anon., 1987; Rovinsky and Wiersma, 1987; Gytarsky *et al.*, 1995).

A gradient approach is applied for forest health assessments. Permanent and temporary sample plots are positioned at various distances along transects laid out in different directions from emission sources. International methods of forest observation and assessment are also applied in some regions of Russia (the Kola Peninsula, Leningrad region), where monitoring is implemented according to international programmes such as the United Nations Economic Commission for Europe (UN-ECE) International Cooperative Programme on the Monitoring and Assessment of Air Pollution Effects in Forests, the Global Environmental Monitoring System, and others (Anon., 1989). Mathematical modelling is used to evaluate the extent of air pollution and its possible effects on forest ecosystems (Gytarsky *et al.*, 1997).

3   Results

The largest sources of atmospheric pollution in Russia are ferrous and non-ferrous metallurgical enterprises and power generators (including thermo-electric power stations). In 1996, total atmospheric industrial emissions (for all pollutants, including SO_2, heavy metals, etc.) by these industries were: 4.75 million tonnes (Mt) for power generation, 3.6 Mt for non-ferrous metallurgical enterprises and 2.53 Mt for ferrous metallurgy (Anon., 1997a).

Fluorine, chlorine, sulphur dioxide (SO_2) and ammonia (NH_3) are the most phytotoxic pollutants for vegetation (Nikolaevsky, 1979). In 1996, there were 2.85 Mt of SO_2 emissions from non-ferrous metallurgical smelters. Power generators and ferrous metallurgy released 2.01 and 0.26 Mt of SO_2 respectively (Anon., 1997a). Emissions of other phytotoxic pollutants were insignificant. Adverse effects of SO_2 on forests are in some cases intensified by heavy metal aerosols released by the enterprises (as in the Kola Peninsula; see Chapters 4 and 5, this volume). However, because of their high deposition velocity and the intensive oxidation and reduction that occurs in the atmosphere, most heavy metal particles are deposited within 10–15 km of the source (Gytarsky *et al.*, 1995). Consequently, the combined impacts of SO_2 and heavy metals are seen in the areas closest to the sources. In contrast, SO_2 can be dispersed over considerable distances and causes severe damage to foliage. Because of this, sulphur dioxide is considered to be the most important phytotoxic pollutant in Russia, and the cause of intense local damage to forests.

Table 3.1. Ministries, agencies and institutions that implement forest monitoring over the territory of the Russian Federation.

Organization	Type of monitoring	Tasks	Area
Federal Forestry Service of the Russian Federation (FFS)			
Department for Forest Control and Protection, Federal Forestry Service		General coordination of activities on forest control and protection from pests, insects and other impact factors	Territory of the Russian Federation
Departments for forest control and protection of the regional Forestry Committees		Coordination of activities on forest control and protection from pests, insects and other impact factors	Within the area of responsibility
Scientific research institutes and analytical centres under FFS	Monitoring of human-induced effects on forest ecosystems including industrial impact, fires, pests and diseases	Collection, analysis and interpretation of data on environmental pollution and damage of forests, caused by anthropogenic activity, pests and pathogenic microorganisms	Permanent and temporary monitoring networks, developed for scientific and experimental purposes
All-Russian Centre for Forest Resources	General forest monitoring	Collection and interpretation of data on the current state of the forest stock in the Russian Federation	Territory of the Russian Federation
Regional forest inventory enterprises	General forest monitoring	Collection and interpretation of data on the current state of the forest stock	Within the area of responsibility
Specialized forest inventory enterprises	Forest pathological monitoring	Forest vitality assessments	Areas with an unfavourable state of forests
Aircraft forest control	Aircraft observations over forests damaged by fires, pests and diseases outbreaks	Opportune detection, control on development and distribution, fire-fighting and actions following outbreaks of insects and pathogenic organisms	Within the area of responsibility
Specialized forest protection enterprises and stations	Forest pathological monitoring	Control over insect and pathogen populations; detection and combating outbreaks; identification of areas of forest decline caused by other impact factors	Within the area of responsibility

State Committee of Russian Federation for Environmental Conservation			
Regional Committees for Environmental Conservation		General coordination of arrangements on forest protection and conservation	Within the area of responsibility
Scientific research institutes and analytical centres	Monitoring of anthropogenic impacts on forest ecosystems	Collection and generalization of data on forest damage caused by anthropogenic activity, including industrial accidents	Permanent and temporary networks developed within the area of responsibility
Federal Service of Russia for Hydrometeorology and Environmental Monitoring (Roshydromet)			
Scientific research institutes and analytical centres under Roshydromet	Monitoring of anthropogenic impacts on forest ecosystems and background monitoring	Collection and interpretation of data on environmental pollution and forest damage caused by anthropogenic impacts including industrial accidents. Collection of data on background environmental contamination	Areas of impact of industrial enterprises, zones of industrial accidents; natural biosphere reserves and protected areas
Russian Academy of Sciences			
Scientific research institutes and analytical centres	Different types of forest monitoring	Collection and interpretation of data on environmental pollution and forest damage caused by anthropogenic activity, pests and pathogens	Permanent and temporary networks, developed for scientific purposes
Ministry of Higher Education of the Russian Federation			
Institutions of higher forestry education	Various types of forest monitoring	Collection and generalization of data on forest damage caused by different types of natural and human-induced impacts	Developed for scientific and educational purposes; permanent and temporary networks

Research has led to the definition of critical levels for SO_2 in the northwestern and central parts of the European territory of Russia and the Ural region. Values vary from 15 to 20 μg SO_2 m^{-3} (Boltneva *et al.*, 1982; Anon., 1984). Studies coordinated by IUFRO have shown that a concentration equal to 25 μg SO_2 m^{-3} is the threshold for the normal growth and development of coniferous forests (Wentzel, 1983). The proposed UN-ECE critical level for SO_2 atmospheric concentrations is 20 μg m^{-3} for forest ecosystems as a whole and 10 μg m^{-3} for highly sensitive components such as sensitive lichen species (UN-ECE, 1993). Critical levels for fluorine of 0.5 μg m^{-3} have been determined for coniferous forest ecosystems in eastern Siberia (Rozhkov and Mikhailova, 1989).

The boundaries of possible areas of industrial impact on forests have been determined and areas (zones) of total, severe, moderate and slight damage to forests revealed on the basis of environmental pollution and forest vitality assessments. Detailed descriptions of forest damage zones have already been given (Boltneva *et al.*, 1982; Karaban and Sysygina, 1988; Aamlid *et al.*, 1995; Gytarsky *et al.*, 1995). Modelled concentrations of SO_2 in the ground air layer averaged over the vegetation period are given in Table 3.2. The concentrations in areas with slight damage are close to the SO_2 critical level for forest ecosystems.

The total area of possible forest decline in Russia may be about 1.3 million hectares, about 0.17% of the total forest area in the country. It includes irreversibly damaged forests, involving over 26,000–65,000 ha (2–5% of the total area of possible forest decline). Severe damage covers 130,000–195,000 ha (10–15%). The remainder consists of zones of moderate and slight damage to forests, covering 30–40% and 40–50% of the total area of possible forest decline, respectively. As the natural regeneration of forest ecosystems takes a long time, particularly in the boreal zone, significant improvements in forest health are not to be expected in the near future in areas subject to long-term industrial impacts.

The development of damage zonation is an integral response of forest ecosystems to long-term industrial impacts (Karaban and Sysygina, 1988; Gytarsky *et al.*, 1995). Zones of forest decline can be clearly linked to the locations of industrial enterprises. Their spatial extent is strictly defined

Table 3.2.   Mean concentrations of sulphur dioxide over zones of different forest damage during vegetation period (from model calculations).

Zone of forest damage	Mean concentration of SO_2 (μg m^{-3})
Total	> 90
Severe	69
Moderate	42
Slight	21
Background area	12

and depends on the amounts and phytotoxicity of the emissions. Damaged areas are therefore distributed unevenly within Russia. They are typical for regions where major emission sources have been functioning for 10 years or more. Such areas include: the north-western (Karelia and Murmansk region) and some of the central (Tula region) parts of the European territory of Russia, the Urals and some Siberian regions (Noril'sk and Irkutsk regions). The greatest damage is currently observed in the Kola Peninsula and at Noril'sk, and these areas have been singled out for special attention in the following chapters. They are among the most highly polluted regions of the Russian Federation.

In the European part of Russia, forest ecosystems are suffering from pollution impacts as a result of the high concentration of industries. In addition, there are many other factors contributing to poor forest health, including recreational activities, cutting, fires, climate and abiotic stresses, insect outbreaks and phytopathogen infestations. These could all be enhancing the susceptibility of forests to industrial pollution. Consequently, it is sometimes difficult to determine the initial causes of specific forest declines. In addition, the resilience of forest ecosystems is determined by their species diversity and natural growth conditions. Forests with only one or a few tree species present and with extreme growth conditions are easily disturbed by unfavourable environmental effects. For example, in the vicinity of the Pechenganikel smelter (Kola Peninsula), the relative occurrence of species decreased by 1.3–1.5 times as a result of industrial pollution. The initial ecosystems were replaced by communities with limited species composition and by industrial barrens. The destructive processes intensified when SO_2 concentrations exceeded 30–35 μg m^{-3} (Vassilieva et al., 1995).

Anthropogenic pressures are very high on forests in urban areas, which are especially important because of their roles for nature conservation and sanitary regulation. In the Moscow region, forest vitality has decreased close to motorways and in the vicinity of local pollution sources in the eastern part of the region.

There may also be clear cases of forest decline in the Ural region. Industrial emissions from the Uphaley nickel enterprise in the southern Urals may be the cause of forest damage over about 20,000 ha in the vicinity of the smelter. However, further surveys are required before the extent of forest decline in the region can be fully assessed.

Large thermoelectric power stations are located throughout the European part of Russia. In particular, the Ryazanskaya Power Station is among the 30 largest European sources of SO_2 emissions (Barrett and Protheroe, 1995). In 1994–1996, the vitality of terrestrial ecosystems in an area adjacent to the Ryazanskaya Power Station was investigated. The surveys did not reveal any indication of forest damage in the vicinity of the station. This appears to be because the technology is properly maintained and because nature conservation arrangements are mandatory. This is confirmed by similar studies performed in the localities of other power stations in different regions of the country (Mozolevskaya, 1996).

4   Conclusions

Local damage to forests occurs in highly polluted regions of the Russian Federation. Generally, these are the localities of large industrial enterprises and areas adjacent to urban agglomerations. Zones with different levels of forest damage have formed in the vicinities of emission sources. They are generally fairly stable in space and time and can be considered as the response of the forest ecosystem to long-term pollution impacts. The identification of various degrees of forest damage provides important information for foresters and local authorities and forms the background for appropriate nature conservation and silvicultural measures aimed at mitigating the adverse effects of industrial pollution on the environment.

The total area of possible industrial damage to forests in the Russian Federation is about 1.3 million hectares, representing 0.17% of the forest area. Significant changes in the area of forest decline are not expected in the near future. Consequently, it will be necessary to apply specific forest regeneration and emission reduction measures over clearly defined areas of forest damage if forest ecosystems are to be preserved for future human generations.

Acknowledgements

This paper is based on the results of long-term observations within the Network of Integral Monitoring of Environmental Pollution and Vegetation State, implemented by the Institute of Global Climate and Ecology, and on the continuous observations on forest health carried out by the All-Russian Institute for Nature Conservation. The authors gratefully acknowledge Vera Kuzmicheva and Nina Surkova, of the Institute of Global Climate and Ecology, for their valuable help during field research and the preparation of this manuscript for publication.

References

Aamlid, D., Tommervik, H., Gytarsky, M., Karaban, R., Venn, K., Rindal, T., Vassilieva, N., Koptsik, G. and Løbersli, E. (1995) Determination of exceedance of critical levels in the border area between Norway and Russia. In: Løbersli, E. and Venn, K. (eds) *Effects of Air Pollutants on Terrestrial Ecosystems in the Border Area Between Norway and Russia. Proceedings of the 2nd Symposium.* Report 1995–8, Directorate for Nature Management, Trondheim, pp. 19–23.

Anonymous (1984) *Temporary Guidelines for Air Pollutant. Critical Levels for Forests of Yasnaya Polyana National Park.* USSR State Committee for Forestry, Moscow (in Russian).

Anonymous (1987) *Temporary Methods Manual for the Determination of the Main Biological Indices of the Integrated Impact Monitoring of Environmental Pollution and Vegetation State.* Ministry of Silviculture of the Russian Federation, Moscow (in Russian).

Anonymous (1989) *Manual on Methodologies and Criteria for Harmonized Sampling, Assessment, Monitoring, and Analysis of the Effects of Air Pollution on Forests.* Programme Coordinating Centre, UNECE International Cooperative Programme on the Assessment and Monitoring of Air Pollution Effects in Forests, Hamburg.

Anonymous (1992) *Sanitary Regulations in the Forests of the Russian Federation.* Ecologia, Moscow (in Russian).

Anonymous (1994) *State Report on the Environmental Situation in the Russian Federation in 1993.* Ministry of Environment of the Russian Federation, Moscow (in Russian).

Anonymous (1997a) *State Report on the Environmental Situation in the Russian Federation in 1996.* State Committee on Environmental Conservation of Russian Federation, Centre for International Projects, Moscow (in Russian).

Anonymous (1997b) *Review of Environmental Pollution in the Russian Federation in 1996.* Roshydromet, Moscow (in Russian).

Barrett, M. and Protheroe, R. (1995) *Sulphur Emissions From Large Point Sources in Europe.* Swedish NGO Secretariat on Acid Rain, Stockholm.

Boltneva, L.I., Ignatiev, A.A., Karaban, R.T., Nazarov, I.M., Rudneva, I.A. and Sysygina, T.I. (1982) Aerosol and gaseous industrial emissions impacts on the northern taiga pine forests. *Ecologia* [Ecology, Sverdlowsk] 4, 37–43 (in Russian).

Gytarsky, M.L., Karaban, R.T., Nazarov, I.M., Sysygina, T.I. and Chemeris, M.V. (1995) Monitoring of forest ecosystems in the Russian Subarctic: effects of industrial pollution. *The Science of the Total Environment* 164, 57–68.

Gytarsky, M.L., Karaban, R.T. and Nazarov, I.M. (1997) On the assessment of sulphur deposition on forests growing over the areas of industrial impact. *Environmental Monitoring and Assessment* 48, 125–137.

Karaban, R.T. and Sysygina, T.I. (eds) (1988) The basics for biological control of environmental pollution. *Proceedings of the Institute of Applied Geophysics, Moscow* 72, 1–172 (in Russian).

Mozolevskaya, E.G. (ed.) (1996) *The Effect of Atmospheric Pollution and other Anthropogenic and Natural Impact Factors on the Decline of Forests of Central and Eastern Europe.* Proceedings of the International Conference (in two volumes). Moscow State Forestry University, Moscow (in Russian).

Mozolevskaya, E.G., Kataev, O.A. and Sokolova, E.S. (1984) *Methods of Forest Health Inspection over the Outbreaks of Bark and Wood Borer Insects and Diseases.* Lesnaya Promyshlennost, Moscow (in Russian).

Nikolaevsky, V.S. (1979) *The Biological Basis for Gas-Resistance of Plant Species.* Nauka, Novosibirsk (in Russian).

Rovinsky, F.Ya. and Wiersma, G.B. (1987) *Procedures and Methods for Integrated Global Background Monitoring of Environmental Pollution.* Environmental Pollution Monitoring and Research Progress Report Series, WMO/TD 178. World Meteorological Office, Geneva.

Rozhkov, A.S. and Mikhailova, T.A. (1989) *The Effect of Fluorine-containing Emissions on Coniferous Trees.* Nauka, Novosibirsk (in Russian).

UN-ECE (1993) *Manual on Methodologies and Criteria for Mapping Critical Levels/Loads and Geographical Areas Where They Are Exceeded*, Text 25/93. Federal Environment Agency, Berlin.

Vassilieva, N., Gytarsky, M. and Karaban, R. (1995) Changes of the dominant species of epiphytic lichens and the ground vegetation cover in the forests subject to 'Pechenganikel' smelter impact. In: Løbersli, E. and Venn, K. (eds) *Effects of Air Pollutants on Terrestrial Ecosystems in the Border Area Between Norway and Russia. Proceedings of the 2nd Symposium*, Report 1995–8. Directorate for Nature Management, Trondheim, pp. 103–111.

Wentzel, K.F. (1983) IUFRO studies on maximal SO_2 emissions standards to protect forests. In: Ulrich, B. and Pankrath, J. (eds) *Effects of Accumulation of Air Pollutants in Forest Ecosystems*. D. Reidel, Dordrecht, pp. 195–302.

The Impacts of Air Pollution on the Northern Taiga Forests of the Kola Peninsula, Russian Federation

4

O. Rigina[1,2] and M.V. Kozlov[3]

[1]Institute of Geography, University of Copenhagen, Copenhagen, Denmark; [2]Institute of the Northern Ecology Problems, Kola Science Centre, Apatity, Russian Federation; [3]Section of Ecology, Department of Biology, University of Turku, Turku, Finland

Copper–nickel smelters in the Kola Peninsula, emitting 350–750 kt of sulphur dioxide annually, have caused visible damage to forests over 39,000 km^2 and forest death over 600–1000 km^2. Individual and population responses of the forest-forming woody plants to pollution are reasonably documented, whereas changes in other components of the forest ecosystems are less well known, and almost no data exist about pollution effects on the functioning of the ecosystems present in the area. The spatial representativeness of the plots used in earlier studies was insufficient for the detailed mapping of forest condition in the region. Remote sensing analysis, supported by ground truth data, has overcome this problem, and has also provided information on the history of pollution impacts in the area. Although the development of predictive models has been delayed by the limited information available about both the emissions and their impacts, forest degradation is expected to continue for some time, even if the emissions are terminated.

1   Introduction

Industrial development in the Kola region started in the 1930s, and there are now numerous plants, mines and open pits situated there (Doiban *et al.*, 1992). The Pechenganikel and Severonikel smelters are the largest sources of emissions in the region (Anon., 1994), and environmental degradation in the Kola Peninsula is amongst the highest in the former Soviet Union (Sokolovsky,

1990), arousing international concern (e.g. Tuovinen *et al.*, 1993; Sivertsen *et al.*, 1994; Anon., 1996). This is especially true of the Pechenganikel smelter, which is situated only *c.* 7 km from the Norwegian border. Concern also exists about transboundary pollution from the Kola Peninsula in Finnish Lapland (Kauhanen and Varmola, 1992; Tikkanen *et al.*, 1992; Tikkanen and Niemelä, 1995). As a result, several national and international research projects on pollution impacts in the region were launched in the early 1990s (Kismul *et al.*, 1992; Kozlov *et al.*, 1993; Tikkanen and Niemelae, 1995; Reimann *et al.*, 1996; Rees and Williams, 1997). Recently, the Kola Peninsula has been one of the main target sites of the Arctic Monitoring and Assessment Program (AMAP, 1997; Nilsson, 1997), the Conservation of Arctic Flora and Fauna project (CAFF, 1996) and the Barents Sea Impact Study (Lange, 1996).

The specific objectives of the present paper are: (i) to give a brief overview of the environmental characteristics of the Kola region relevant to the understanding of pollution impacts there; (ii) to outline the current knowledge on how pollution affects forest vegetation in the region; (iii) to describe how forest health has been assessed at both local and regional scales; and (iv) to discuss the past, present and future of the Kola forests.

2   Abiotic environment

2.1   Topography and climate

The Kola Peninsula (*c.* 100,000 km^2) is part of the Murmansk county and is located in the north-west extremity of Russia (Fig. 4.1). The highest mountains in the area, Khibiny (1191 m) and the Lovozerskie tundras (1126 m), are formed from alkaline rocks and situated in the central part of the peninsula. Other massifs (Monche-, Volchji-, Salnye-, Iolgi-tundra and Kolvitskie tundras) are formed from basic and ultra-basic rocks. The hydrological net is dense, with numerous lakes dominated by Imandra (816 km^2), Notozero (745 km^2) and Umbozero (313 km^2). Rivers (the largest are Ponoy, Iokanga, Varzuga and Tuloma) are mostly controlled by snow melt.

The climate varies from maritime along the coast to moderately continental in the central part of the peninsula. Mean annual temperature is 0°C. Winter lasts for *c.* 7 months, but is moderated by the effects of the Gulf stream: the minimum temperatures vary from −33°C by the sea to −40°C in continental regions. The cool summer lasts for *c.* 3 months; the sum of effective temperatures (over +5°C) is 780–800 degree-days. The frost-free period exceeds 100 days along the coasts and ranges from 50 to 100 days inland. The number of days with snow cover varies from 180 to 200 in the lowlands to 220 in the mountains. Annual precipitation ranges from 500 to 700 mm, reaching 1000 mm in the mountains. Northern and north-eastern winds are most frequent in summer; southern and south-western winds occur mainly in other

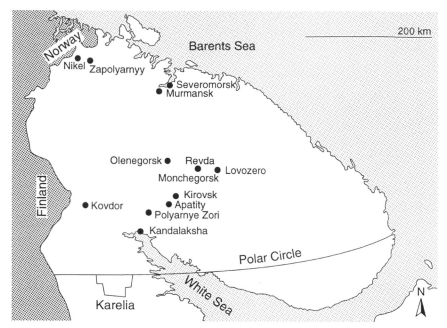

Fig. 4.1.   Principal sites of human activities on the Kola Peninsula: Murmansk,
naval shipyards, nuclear vessels; Zapolyarnyy, open pit Ni mine, Cu–Ni ore
roasting; Monchegorsk, the Severonikel smelter; Nikel, the Pechenganikel smelter;
Lovozero, Revda, mine (U etc.); Kirovsk, open pit apatite mine, apatite processing;
Apatity, coal-fired power station, nepheline waste, apatite processing; Severomorsk,
military base, nuclear vessels; Olenegorsk, open pit iron mine, mica mine;
Kandalaksha, Al smelter; Polyarnye Zori, nuclear power plant.

seasons. However, the wind pattern is very heterogeneous across the region
and is mostly conditioned by local topography and hydrology.

2.2   Emission rates and emission sources

The Murmansk county is one of the most industrially developed and urbanized
areas in the Russian North (Doiban *et al.*, 1992). About 170 enterprises have
emitted annually 600–900 kt of pollutants into the atmosphere over the last
decade (Baklanov, 1994). In addition, 176 kt of pollutants are generated by
traffic (Murmansk Regional State Committee on Ecology, 1996). The main
pollution sources are the copper–nickel smelters of Pechenganikel in Nikel and
Severonikel in Monchegorsk, the Kandalaksha aluminium plant and some
enterprises in Murmansk.

Emissions in the region in 1995 were dominated by SO_2 (451 kt). Other
pollutants were CO (112 kt), NO_x (25 kt), hydrocarbons (17 kt), nickel

(1.6 kt), copper (0.9 kt), cobalt (0.05 kt), fluorine (0.8 kt), phenol (0.001 kt) and formaldehyde (0.03 kt) (Murmansk Regional State Committee on Nature Protection, 1997). Earlier (1987–1991), annual emissions were higher (513 kt of SO_2, 30 kt of NO_x, 3.7 kt of nickel, 2.6 kt of copper, 1.6 kt of fluorine: Baklanov, 1994). The recent decrease in emissions from both the Pechenganikel and Severonikel smelters (Fig. 4.2) is mostly related to a reduction in metal production.

The Pechenganikel smelter was established in 1933 in Nikel as a joint venture between Finland and Canada. In 1959, a new ore-roasting workshop was established in Zapolyarnyy. The smelter used indigenous ores (6.5% sulphur) until 1973, and has used ores from Noril'sk (30% S) since then.

The Severonikel smelter near Monchegorsk has been in operation since 1938 for processing the sulphide copper–nickel ores (1.15% S) from the local Nittis-Kumuzhje deposit. In 1946–1947, ores from the Pechenga district (6.5% S) were used. Utilization of sulphur from the metallurgical gases and production of sulphuric acid was started in 1968. As an immediate result, emissions of SO_2 dropped from c. 135 to 70 kt (Alexeyev, 1993). In 1969, the smelter started to use Noril'sk ores, and emission levels increased dramatically.

The Kandalaksha aluminium plant (established in 1951) is the main source of emissions of hydrogen fluoride and insoluble fluorides in the region. It emitted 50 kt of pollutants annually during 1987–1991 (Ratkin, 1993).

The above emission figures were estimated by the mass balance method from the amount of raw materials used and the products obtained. Although the accuracy of this method is quite high, and the error for sulphur is probably

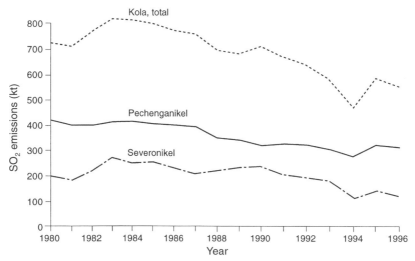

Fig. 4.2. SO_2 emission rate from Severonikel, Pechenganikel and total Kola (data from the Murmansk Regional Committee of Nature Protection).

less than 10% (Tuovinen *et al.*, 1993), estimates for other components are very approximate (A.A. Baklanov and J. Kelley, personal communication). Changes in the calculation methods in 1989 resulted in changes in the figures by a factor of 10 for some components (Baklanov and Rodyushkina, 1993).

2.3   Monitoring of atmospheric pollutants

The monitoring system in the Kola Peninsula consists of two levels: (i) regional (within the framework of the European Monitoring and Evaluation Programme); and (ii) local: at major settlements and industrial sites (Table 4.1 and Fig. 4.3). Depending on the station, information on SO_2 varies from continuous data to daily means; for heavy metals, the daily means are usually reported.

Ambient concentrations

The Severonikel smelter has ten stacks with a release height over 100 m; the highest one is 200 m. Pechenganikel has three stacks of 150–160 m height in Nikel and one 100 m stack in Zapolyarnyy. Emissions from the high stacks can be transported as far as 2000 km from the smelters (Jaffe *et al.*, 1995). The low sources (release height below 50 m) mainly contribute to the contamination of areas 20–30 km from the smelter (Baklanov and Sivertsen, 1994). Contamination by these sources is exacerbated by frequent temperature inversions (Kryuchkov and Makarova, 1989; Baklanov *et al.*, 1994).

The largest air pollution problems in the region are episodic, i.e. events of high ambient concentrations lasting a few hours to 2 days (Sivertsen *et al.*,

Table 4.1.   The monitoring system in the Kola peninsula.

Site	Station	Operator	Service period	Site	Station	Operator	Service period
The Severonikel impact zone				The Pechenganikel impact zone			
1	Monchegorsk	Hydromet	Permanent	7	Nikel	Hydromet/ NILU	Permanent
2	'Severonikel'	'Severonikel'	Permanent				
3	Chuna	Lapland Reserve	1985–1990	8	Zapolyarnyy	Hydromet	Permanent
				9	Kobbfoss	NILU	
4	Khibiny	INEP	1991–1993	10	Maayarvi	INEP/ Hydromet	Since 1990
5	Kislaya	INEP	1991–1994				
6	Profilaktorij	INEP	1991–1995	11	SOV1	INEP/NILU	1990–1994
Regional monitoring				12	SOV3	INEP/NILU	1990–1995
				13	Holmfoss	NILU	Since 1978
17	Karpdalen	EMEP	Permanent	14	Viksjofjell	NILU	Since 1988
18	Kirkenes	EMEP	Permanent	15	Svanvik	NILU	Since 1974
19	Janiskoski	EMEP	Permanent	16	Noatun	NILU	1988–1991

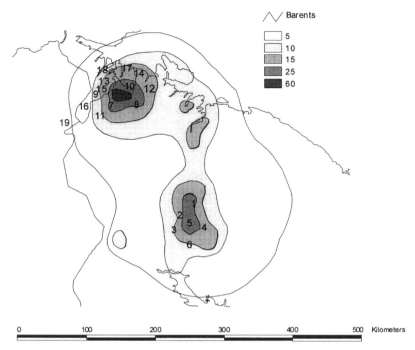

Fig. 4.3. Stations for monitoring air quality (the numbers are the same as in Table 4.1) and mean annual concentration of SO_2 (μg m^{-3}) in the Kola Peninsula in 1995 (after A.A. Baklanov, unpublished observations).

1994). Episodes occur during 3–5% of days in winter and 1–2% in summer (Baklanov and Sivertsen, 1994). During episodes, hourly concentrations of SO_2 exceed the seasonal and monthly means by a factor of 10–25 (Baklanov and Sivertsen, 1994) and can reach 1200 μg m^{-3} in the impact zone of the Severonikel smelter (Baklanov and Rodyushkina, 1993) and 2000–2500 μg m^{-3} in the Pechenganikel impact zone (Baklanov et al., 1994; Sivertsen and Bekkerstad, 1996). Concentration fields (averaged for a specific period) can be predicted by mathematical models and verified by monitoring data (Fig. 4.3). In 1992, mean annual SO_2 concentrations exceeded 25 μg m^{-3} over c. 1200 km^2, and 10 μg m^{-3} over c. 4800 km^2 around the Pechenganikel smelter (Sivertsen et al., 1994).

Ambient concentrations of heavy metals, measured during SO_2 episodes, exceeded the background levels by factors of 10–20 for nickel and copper, 5–10 for cobalt, cadmium and arsenic, and 1–2 for zinc and chromium (Sivertsen et al., 1994). Metals in the fine fraction of dust (particle size < 1 μm) such as cadmium, lead, zinc and arsenic, are transported longer distances than metals mainly associated with the coarse fraction, such as iron, nickel, copper, manganese and cobalt (Sivertsen et al., 1994).

Deposition of pollutants

The deposition patterns of heavy metals and sulphates in the Kola Peninsula are similar, although with different spatial scales due to the effects of gravitational settling (Hagner and Rigina, 1998). The maxima are close to the copper–nickel smelters; the spatial distribution of pollutants around the emission sources are determined by the meteorology of the region and the release characteristics (Rigina and Baklanov, 1998). The dry fraction constitutes up to 80% of the total sulphur deposition near the Pechenganikel smelter (Sivertsen *et al.*, 1994), but beyond distances of *c.* 50 km, the proportion of wet deposition may increase to up to 50% of the total deposition (Baklanov, personal communication). The regional level of sulphur deposition on the Kola Peninsula in 1979–1983, estimated on the basis of snow sampling, was $0.7–1.0$ g S m^{-2} year^{-1} (Kryuchkov and Makarova, 1989). This level was twice as high as in 1966 (data of Gurevich, 1966), presumably due to the processing of high-sulphur ores from Noril'sk since 1969. About 45% of the sulphur deposition is due to transboundary transport (Kryuchkov and Makarova, 1989). Although it is believed that the role of nitrogen compounds will increase in future (Brodin and Kuylenstierna, 1992), recent acidification in the Kola Peninsula is mostly related to sulphur deposition (Rigina and Baklanov, 1998).

Deposition of nickel and copper exceeded the background values by a factor of ten or more within 10–30 km of the smelters (Evseev and Krasovskaya, 1990; Sivertsen *et al.*, 1994). Annual nickel deposition in the impact zones of the smelters ranged from 50 to 100 mg m^{-2} in the late 1980s to early 1990s (Evseev and Krasovskaya, 1990; Baklanov, 1994; Jaffe *et al.*, 1995). Although metal deposition is generally confined to smaller areas (within 40 km radius, Tikkanen and Niemelä, 1995) than sulphur deposition, metal concentrations in lichens in the late 1980s exceeded the background level by a factor of 2–5 over more than half of the Murmansk region (Evseev and Krasovskaya, 1990).

3 Past effects of pollution on forest vegetation

3.1 Extent of forests in the Kola Peninsula

The total area covered by the forest biome in Murmansk county in 1988 was estimated as 99,920 km^2, including 22,340 km^2 of pine, 16,100 km^2 of spruce forests, and 13,460 km^2 of birch (Tsvetkov, 1990b; Anon., 1991). Forests dominated by Scots pine (*Pinus sylvestris* L.), generally associated with dry sandy podzol soils with low content of nutrients, are more abundant in the south-eastern part of the region. Norway spruce (*Picea abies* (L.) Karst.) is mostly confined to wet illuvial podzols with higher nutrient contents; spruce forests dominate in central and south-eastern parts of the Kola Peninsula

(Tsvetkov, 1991). On mountains, conifers reach altitudes of 350–380 m a.s.l., and the tree-line is formed by mountain birch, *Betula pubescens* subsp. *czerepanovi* (Orlova) Hämet-Ahti (Tsvetkov, 1990b).

Although the standing biomass of the forests of the Kola Peninsula is quite low, from 25 to 170 m^3 ha^{-1}, fellings have been intensive (Tsvetkov, 1990b; Alexeyev, 1993). The over-exploitation, along with the lack of reforestation measures, has resulted in the replacement of coniferous forests by birch-dominated communities over large areas (Tsvetkov, 1990b).

3.2   Investigations of the pollution effects on the Kola Peninsula

Three basic approaches have been applied to reveal the local effects of emissions on forest health: (i) comparison between study sites which differ in pollution loads; (ii) ground survey of the impact area followed by subdivision into several zones which differ in the level of environmental degradation; and (iii) remote sensing with ground verification of the satellite data. Approaches (i) and (ii) presume that the temporal impact of continued pollution is analogous to moving closer to the smelter, i.e. the zone of maximum damage will gradually expand (Kozlov and Haukioja, 1995). Some existing data sets extend to 15 years, but a truly historical approach, involving the re-investigation of sites described in detail before the smelters started to operate, has rarely been applied (Kryuchkov and Syroid, 1984; Koroleva, 1993).

Several research groups have established study plots around the Severonikel smelter, and some of these plots have been monitored for several years (for an earlier history of research in this area, see Alexeyev, 1993). Most studies provide detailed descriptions of the vegetation type and stand composition, the chemistry of soils and soil water and the tree health – damage class (Doncheva, 1978; Kryuchkov and Makarova, 1989; Alexeyev, 1990b; Norin and Yarmishko, 1990; Koroleva, 1993; Yarmishko, 1993, 1997; Lukina and Nikonov, 1996). The most detailed studies (Lukina and Nikonov, 1993, 1996; Nikonov and Lukina, 1994) have taken into account the spatial variability of forest ecosystems.

Plots aimed at monitoring local effects are mostly located to the south of the Severonikel smelter, i.e. in the prevailing wind direction during the growth season. Only a few studies (Fedorkov, 1992; Koroleva, 1993; Yarmishko, 1993; Zvereva *et al.*, 1997a) have involved true spatial replication of pollution loads by arranging plots along two gradients, to the north and south of the smelter, or by spreading study sites evenly over the impact zone (Barkan *et al.*, 1993b). Long-distance impacts of emissions from both the Pechenganikel and Severonikel smelters have been evaluated along transects crossing northern Finland by the Lapland Forest Damage Project (Tikkanen and Niemelä, 1995).

Health indices developed by Alexeyev (1989, 1990a) have enabled the distinction between healthy, damaged, strongly damaged and completely destroyed stands. Together with the assessment of the health of other components of forest ecosystems, including both those that are more sensitive (epiphytic lichens) and those that are less sensitive to pollution (some components of the underground flora) than conifers, the indices enable the evaluation of the health status of the ecosystem (Alexeyev, 1993, 1995). However, in practical forestry, only the evaluation of stand quality has been conducted (Bobrinsky, 1993; Vassilieva, 1993). In environmental studies, the assessment of forest health has frequently been restricted to the estimation of the longevity of needles on conifers (Kryuchkov, 1993; Kozlov and Haukioja, 1995). Recently, fluctuating asymmetry was introduced as a tool for the evaluation of the environmental stress imposed on deciduous plants (Kozlov et al., 1996; Zvereva et al., 1997a) and conifers (Kozlov and Niemelä, 1999).

Following the 'landscape degradation' classification system of Doncheva (1978), most studies have defined three to six zones on the basis of the deposited pollutants and/or the state of forest ecosystems as revealed by ground surveys (Kryuchkov and Makarova, 1989; Norin and Yarmishko, 1990; Armand et al., 1991; Tsvetkov, 1993; Lukina and Nikonov, 1996). However, the definition of the zone boundaries remains controversial and subjective. In comparison with ground survey methods alone, satellite remote sensing has several advantages (Nilsson and Duinker, 1988). Remote sensing on the Kola Peninsula started in the 1960s with the American satellite series Corona (2 m resolution). Later, Landsat-MSS (80 m resolution, from 1972) imagery became available, followed by Landsat-TM (30 m resolution, from 1982), Russian COSMOS (2 and 10 m resolutions, from 1979) and RESURS-01 (45 [MSU-E] and 175 m [MSU-SK] resolutions, from 1989), SPOT (10 and 20 m resolutions, from 1986) and IRS-1C (5 m resolution, from 1996). Images from Russian military satellites are expected on the market in the future.

Multi-spectral changes detected in the forests around the Severonikel smelter during 1978–1992 were spatially correlated with the modelled long-term average SO_2 concentration levels (Fig. 4.4) (Hagner and Rigina, 1998). Similarly, Mikkola (1996) found a positive correlation between SO_2 levels and thermal radiation (Landsat-TM, band 6) around the Pechenganikel smelter. The spectral reflectance of a common moss species, *Hylocomium splendens* (Hedw.) B.S.G., is affected by the concentrations of metal pollutants accumulated in its tissues, and can thus be used for remote sensing of environmental contamination (Buznikov et al., 1995b). Tømmervik et al. (1995) and Høgda et al. (1995) also used lichens (*Cladina* spp.) as a remotely sensed bioindicator in the Pechenganikel impact zone. Spectral signatures of damaged and healthy trees have also been distinguished (Buznikov et al., 1995a; Høgda et al., 1995; Kravtsova et al., 1995; Solheim et al., 1995; Tømmervik et al., 1995; Mikkola, 1996; Rees and Williams, 1997; Hagner and Rigina, 1998). Remote sensing has successfully detected forest damage around both the Pechenganikel (Roberts et al., 1994; Høgda et al., 1995; Johansen et al., 1995; Tømmervik et al., 1995) and Severonikel (Buznikov et al., 1995a;

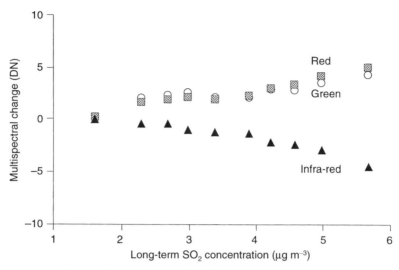

Fig. 4.4. Multispectral change, DN (digital number) for Landsat MSS bands green, red and near-IR in forest vegetation versus long-term mean SO_2 ($\mu g\ m^{-3}$) concentrations ($Y = b_0 + b_1 \ln(SO_2)$, $P = 0.000$, $r^2 = 0.788$, 0.932, 0.922 respectively) in air in the Severonikel smelter impact zone for the period 1978–1992 (after Hagner and Rigina, 1998).

Kravtsova *et al.*, 1995; Solheim *et al.*, 1995; Mikkola, 1996; Rees and Williams, 1997; Hagner and Rigina, 1998) smelters. It has been possible to stratify six zones of pollution impact, from industrial barrens to reference areas with no pollution-related damage (Tømmervik *et al.*, 1995; Mikkola, 1996).

Some attempts have been made to model forest decline due to air pollution at the level of individual trees, compartments or regions (Alekseev, 1993, 1995; Tarko *et al.*, 1995; Rigina *et al.*, 1999). However, most of the estimations provided in these publications are very coarse and should be used in combination with data from ground and/or remote surveys.

3.3   Sensitivity of forest vegetation to pollution

In the tundra and taiga biomes, the first sign of deterioration caused by SO_2 and heavy metal pollution is the decline of lichens (Kryuchkov and Makarova, 1989; Pastukhova *et al.*, 1992). Consequently, in the forest ecosystems of the Kola Peninsula, epiphytic lichens have been used as the most sensitive bioindicators of contamination (Ablaeva, 1978; Gorshkov, 1993; Kryuchkov, 1993). For the protection of lichens, mosses and sensitive tree species, the annual average SO_2 concentration should not exceed 20 $\mu g\ m^{-3}$ (UN-ECE, 1988), the daily means should remain below 70 $\mu g\ m^{-3}$ and hourly averages should not exceed 150 $\mu g\ m^{-3}$. Lower SO_2 concentrations may also result in plant damage (Sutinen and Koivisto, 1992), especially in combination with other stressors such as low temperatures (Scherbatyuk, 1988). Consequently, an annual critical level of 15 µg m^{-3} has been suggested for northern forests (UN-ECE, 1993). For the most sensitive elements, such as the cyanobacteria and lichens, the critical level is 10 µg m^{-3} (UN-ECE, 1993). This level is consistent with suggestions by Kryuchkov (1993) who concluded that both pine and spruce forests of the Kola Peninsula can survive SO_2 concentrations of 5–9 µg m^{-3} for an unlimited time. However, some researchers have suggested lower values: Tsvetkov (1993) argued for a critical level of 7 µg m^{-3}, and Manninen and Huttunen (1997) proposed a critical level of 5 µg m^{-3} for both the growth season and annual mean.

The critical load for sulphur in the ecosystems of the Kola Peninsula has been estimated to be less than 0.3 µg S m^{-2} year^{-1} (Downing et al., 1993) at the regional scale. However, Vasilenko et al. (1992) estimated it as 0.3 µg S m^{-2} in the south-western part of the region, 0.2 µg S m^{-2} around Severonikel and 0.17–0.18 µg S m^{-2} on the Barents Sea coast, including the Pechenganikel area and surroundings of Murmansk. Koptsik and Koptsik (1995) produced a map of critical loads for soils, with values ranging from 0.2 µg S m^{-2} in the eastern parts of the peninsula to 0.8 µg S m^{-2} in the mountains (Khibiny, Lovozerskie and Monche tundras).

Critical levels of heavy metals in soils (as adopted in Russia) are 40 µg g^{-1} for copper and 45 µg g^{-1} for nickel (Evdokimova, 1993). Consistent with this level, the forest vegetation south of the Severonikel smelter underwent changes after soil concentrations of soluble nickel rose above 3–4 µg g^{-1}, equivalent to 49 µg g^{-1} of total nickel (Armand et al., 1991).

3.4   Current state of the Kola ecosystems

Unpolluted areas are confined to the eastern parts of the Kola Peninsula (Evseev and Krasovskaya, 1990), and undisturbed primary forests are only found in some remote areas such as the Ponoy valley (see Neshataev and Neshataeva, 1993). Thus, most of the 'control' plots used in past environmental studies have been subject to human-induced disturbance. However, these disturbances are very small in comparison with the effects of the high pollution loads.

Undamaged forests. Detailed descriptions of the background state of the Kola forests were provided by Vatkovskij (1990) and Alexeyev (1995). Unpolluted forests are not always healthy. A 'normal' value of the index of the forest state (which has the lowest value of 0 and highest of 1) in the Kola Peninsula was 0.84–0.99 for young and mature forests, but only 0.65–0.70 for over-mature stands (Alexeyev, 1995).

Non-visible damage. This occurs at mean SO_2 concentrations ranging from 2 to 8 µg m^{-3} (Tikkanen and Niemelä, 1995), and/or concentrations of copper and nickel in the humus (A_0) soil horizon about twice as high as the background level. Microscopic damage to Scots pine needles is frequently accompanied by chlorosis of needle stomata (Raitio, 1992) and premature needle loss (Kryuchkov, 1993; Yarmishko, 1997), but radial increment remains unaffected (Alekseev, 1993). Both the species composition of epiphytic lichens (Gorshkov, 1993) and the physiological functions of lichens (Plakunova *et al.*, 1983; Tarhanen *et al.*, 1996) are modified. The continuous moss cover is dominated by long-living species, which generally have well-developed photosynthetic (green) parts of the gametophyte (Andreeva, 1989).

Slight (visible) damage. This corresponds to average SO_2 concentrations of 8–15 µg m^{-3} (Tikkanen and Niemelä, 1995); concentrations of copper and nickel in the A_0 horizon exceed the background level two- to threefold (Barkan *et al.*, 1993b). Surface waxes on pine needles are eroded, and visible symptoms of injury, such as tip necroses, affect up to 10% of pine and spruce needles (Kryuchkov, 1993). The longevity of spruce needles ranges from 5 to 10 years, with needle mass reduced to about half of the level typical for unpolluted forests; spruce trees with dead upper canopies first appear at this stage of damage (Kryuchkov, 1993). The loss of *c.* 60% of needle biomass corresponds to a rapid decline in the radial increment of Scots pine (Alekseev, 1993). Cone production by Scots pine declines to about half of the background level (Stavrova, 1990). Ground lichens and mosses deteriorate; in many moss species, only the current-year part of the gametophyte remains green (Andreeva, 1989). The proportion of shrubs in the ground vegetation decreases, whereas the abundance of *Deschampsia flexuosa* (L.) Trin. in spruce forests increases. Barren spots (with no ground layer vegetation) cover less than 5% of the forest area (Rigina *et al.*, 1999).

Moderate damage. This results from average SO_2 concentrations of 15–40 µg m^{-3} (Tikkanen and Niemelä, 1995); concentrations of copper and nickel in the A_0 horizon exceed the background level by 3–20 times (Barkan *et al.*, 1993b). Spruce mortality reaches 30–40%; living trees have lost up to 75% of their needle biomass and are frequently flag-shaped and/or have dead upper canopies (Kryuchkov, 1993). Visual damage on needles is quite prominent; regeneration of conifers is poor, both in terms of pollen and seed vitality (Stavrova, 1990; Fedorkov, 1992) and seedling establishment (Stavrova, 1990). The abundance of bilberry (*Vaccinium myrtillus* L.) and cowberry (*Vaccinium vitis-idaea* L.) decreases and the more resistant crowberry (*Empetrum nigrum* L.) and *D. flexuosa* become dominant (Rigina *et al.*, 1999). The moss flora may be more diverse than in unpolluted forests, but it is dominated by short-living (1–5 years) species (Andreeva, 1989). Bare ground exceeds 5% of the forest area (Rigina *et al.*, 1999).

Severe damage. Average SO_2 concentrations exceed 40 µg m^{-3} (Tikkanen and Niemelä, 1995); concentrations of copper and nickel in the A_0 horizon exceed the background level by 20–50 times (Barkan *et al.*, 1993b). Most of the coniferous trees are dead (Kryuchkov, 1993), and birches are stressed (Kozlov *et al.*, 1996), with prominent foliar necroses later in the growing season. Dieback of woody plants results in twice as much fuel wood as normal (Chekrizov, 1987) which, when combined with the proximity to the industrial sites, increases both the frequency and severity of fires (Tsvetkov, 1990a, 1991) and promotes the transformation of forests into industrial barrens. The ground vegetation cover decreases to below 50% and is dominated by crowberry. Mosses are represented by patchy sinusia of pioneer species on eroded soils (Andreeva, 1989).

Decline. Average SO_2 concentrations exceed 100 µg m^{-3} (M. Kozlov and E. Haukioja, unpublished data); concentrations of copper and nickel in the A_0 horizon exceed the background level by 50–80 times (Barkan *et al.*, 1993b). Denuded landscapes adjacent to the largest sources of SO_2 and heavy metals are often called industrial barrens (Doncheva, 1978; Kryuchkov and Makarova, 1989; Tsvetkov, 1991; Kozlov and Haukioja, 1995). These landscapes are formed on acidic, metal-contaminated soils, devoid of large (> 3 m tall) trees and bushes, with a vegetation cover less than 10%, and a relict soil cover less than 20%. The illuvial soil horizon is exposed over more than 80% of the surface and suffers from intensive erosion (Kozlov *et al.*, 1999). Nearly all coniferous trees are dead; birches grow as low bushes, with dead upper canopies.

3.5   Expansion of the damaged areas

Forests up to 6 km from the Severonikel smelter were dead by 1946 (Berlin, 1993). Between 1955 and 1963, the area of severe mortality was restricted to within 6–7 km of the smelter. However, by 1973, the area without trees was estimated to cover *c.* 50 km^2 (Doncheva, 1978), and this area had doubled by 1983–1987 (Doncheva *et al.*, 1992). In recent decades, the damaged area around the Severonikel smelter has steadily expanded (Pozniakov, 1993; Tsvetkov, 1993; Mikkola, 1996). However, there is some doubt about the degree of expansion of the damaged area (Kozlov and Haukioja, 1995). The exceptionally large extent of the damage identified in 1992 may be the result of either an accidental pollution discharge (for example, on 12 April 1992, see Baklanov *et al.*, 1993) or an exceptionally cold spring, or a combination of the two.

Although guidelines for the delimitation of damage zones vary, there is good agreement over zonal borders, at least south of the Severonikel smelter (Rigina *et al.*, 1999): undamaged, beyond 60 km; non-visible damage, 35–60 km; slight damage, 20–35 km; moderate damage, 12–20 km; severe damage, 7–12 km; forest decline, within 7 km distance. Satellite images taken in different years (Solheim *et al.*, 1995; Mikkola, 1996; Rees and Williams, 1997; Hagner and Rigina, 1998) indicate that the zones of forest death and damage have expanded to the north of the Severonikel smelter more than to the south. However, a brief visit to one of the areas of 'forest death' described by Mikkola and Ritari (1992) in 1996 revealed young (12–17 years old) pine forest recovering after fire (M. Kozlov, unpublished data), emphasizing the need for ground surveys to verify the spatial pattern of forest damage as revealed by satellite imagery.

Around the Pechenganikel smelter, the area damaged by pollution (with an annual mean SO_2 concentration exceeding $50 \mu g \ m^{-3}$) increased from $55 \ km^2$ in 1973 to $760 \ km^2$ in 1988. The annual increase in the extent of the affected area was 5 km between 1973 and 1979, 4 km during 1979–1985 and 1 km during 1985–1988 (Høgda and Tømmervik, 1996). Recently, extensive damage to vegetation, mostly peat moors and birch and pine forests, was recorded over an area of $2000 \ km^2$. Slight and moderate damage, mostly affecting lichens, occurred in an area of $1900 \ km^2$, and a further $770 \ km^2$ are subjected to minor episodic damage (Tømmervik *et al.*, 1995). As a result of the recent decline in emissions (Fig. 4.2), the expansion of damaged areas around both the Pechenganikel and the Severonikel smelters seems to have stopped (Høgda and Tømmervik, 1996; Murmansk Regional State Committee on Nature Protection, 1997; Rigina *et al.*, 1999). The areas around Kandalaksha that are the most heavily damaged (Georgievskii, 1990) seem to correspond to the zone of moderate damage around the Severonikel (see Kryuchkov, 1993). The total area affected by emissions from the aluminium plant was *c.* $1300 \ km^2$ in 1980 (Georgievskii, 1990). Zones of non-visible and slight damage around the Severonikel and Kandalaksha plants had merged by 1986 (Kryuchkov and Makarova, 1989).

The critical load approach (Downing *et al.*, 1993) has been applied to evaluate the detrimental effects of pollution on the ecosystems of the Kola Peninsula (Moiseenko, 1994; Aamlid *et al.*, 1995; Koptsik and Koptsik, 1995; Rigina and Baklanov, 1998). Contamination exceeds critical limits over about half of the Kola Peninsula. Regional maps of sulphur deposition indicate that sulphur critical loads ($0.3 \ g \ m^{-2} \ year^{-1}$, according to Downing *et al.*, 1993) were exceeded in 1990/91 over an area of $150,000 \ km^2$, including $99,000 \ km^2$ of Russia, $32,000 \ km^2$ of Finland and $19,000 \ km^2$ of Norway (Tuovinen *et al.*, 1993).

4   Information needs and construction of predictive models

Even the highest levels of SO_2 and heavy metals recorded around the Kola smelters do not immediately kill all the exposed trees. Acute leaf damage may occur at SO_2 levels of approximately $1000 \mu g \ m^{-3}$ (Baklanov and Sivertsen, 1994). The degradation of forest ecosystems exposed to pollution takes several decades (Kryuchkov and Syroid, 1987) and it is important to determine the future of moderately and slightly polluted ecosystems under continued pollution exposure. However, information is currently insufficient to make sound predictions.

4.1   Information needs for pollutants

Measurements of airborne pollutants during the international Monchegorsk field campaigns in the summers of 1992–1994 revealed the presence of numerous poorly investigated compounds such as trace elements, Pb, As, Cr, Co, Zn, Cd, Al, V, phenols, formaldehyde, benzpyrenes, dioxin and other organochlorides (Baklanov *et al.*, 1997). Field measurements also revealed episodes with high concentrations of O_3 (Tuovinen and Laurila, 1993), although the origin of these episodes remains unknown.

The investigation of pollution impacts on forest ecosystems in the Kola Peninsula requires:

1.   Probability analyses of trajectories of individual pollutants;
2.   Mapping of trace elements, organic compounds and ozone;
3.   Additional inventory of the emission sources of less-studied pollutants;
4.   A study of aerosol size distribution; and
5.   Studies of the temporal variability of emissions and depositions.

Some of the toxic compounds may undergo fast degradation, whereas others, especially heavy metals, are accumulated in soils and persist for decades. Although some experiments on metal leaching from contaminated soils have been conducted (e.g. Barkan *et al.*, 1993a; Evdokimova, 1995), they have not yet resulted in the input–output balance models that are urgently needed to predict the dynamics of the impact zones under different emission scenarios.

4.2   Information needs for direct impacts

Individual responses of trees (Scots pine, Norway spruce, and mountain birch) to the emissions of the copper–nickel smelters are reasonably well documented (Alexeyev, 1990b; Norin and Yarmishko, 1990; Lukina and Nikonov, 1993). This enables short-term predictions on the state of individual trees or stands, but these predictions are only valid for periods of gradual change in the affected ecosystems. After passing some (unknown) threshold level, the ecosystem may undergo a catastrophic, unpredictable change. Although there have been some attempts to identify the threshold levels of pollutants (Armand *et al.*, 1991), the available information is still very descriptive. Very little information is available on the reversibility of the environmental changes that may occur.

Even less is known about changes in other components of forest ecosystems. Of the insects, only the population dynamics of the willow-feeding leaf beetle, *Melasoma lapponica* L., have been studied in relation to environmental contamination (Zvereva *et al.*, 1997b). Density records along pollution gradients were published for several other herbivores, including an autumnal moth (*Epirrita autumnata* Bkh.), which is an important defoliator of subarctic birch forests (Ruohomäki *et al.*, 1996). Data on forest pests are scarce (Sharapa and Pisareva, 1987). Impacts on microbes, earthworms, ants and the vast majority of invertebrates are virtually unknown (Kozlov and Haukioja, 1995). Thus, it is practically impossible to create a model describing the impacts on even the most important links between the principal components of forest ecosystems.

The field data which have been collected so far do not allow the distinction between the individual effects of different pollutants. This means that predictions derived from these data are valid only if the relative amounts of the different pollutants remain constant. The situation may be clarified by additional controlled experiments (cf. Neuvonen, 1993).

4.3   Information needs for indirect effects

Pollution also affects forest biota through changes in temperature and other climatic characteristics of the environment, at both regional and local scales. Microclimate changes recorded in polluted habitats may: (i) directly affect seasonal development and performance of plants and animals; and (ii) lead to combined effects including damage by pollution-induced changes in sensitivity to extreme temperatures or temperature-related changes in sensitivity to pollutants (Kozlov and Haukioja, 1997).

Increased soil temperatures of barren soils during the growth season may decrease root hardening (Sakai and Otsuka, 1970) which, in combination with a pollution-induced decrease in cold-hardiness (Sutinen *et al.*, 1996), increases the probability of death from freezing injury in severely polluted habitats which are colder in winter time than healthy forests (Kozlov and Haukioja, 1997). Conversely, temperature may affect plant sensitivity to pollution (Norby and Kozlowski, 1981; Mortensen and Nilsen, 1992; Mortensen, 1993). Thus, the observed forest decline in polluted habitats, although presumably triggered by pollution, may not be attributable to the toxic effects of pollutants alone: initial pollution-induced forest disturbance, through secondary effects, may enhance further disturbance in a positive feedback cycle (Perkins *et al.*, 1987).

4.4   Desirable reductions in pollution

First estimates (Kryuchkov, 1988; Kryuchkov and Makarova, 1989) recommended a reduction in the annual emissions of sulphur dioxide by the Severonikel smelter to 23 kt, by the Pechenganikel smelter to 10 kt, and by the Kandalaksha aluminium plant to 3–5 kt. These measures were expected to prevent further damage to forest ecosystems. In another publication, Kryuchkov (1989) suggested 20 kt as the maximum permissible emission from all sources in the Murmansk region. The total emissions of metals need to be reduced to less than 0.1–0.2 kt, as was the case in the mid-1960s (Kryuchkov, 1988; Kryuchkov and Makarova, 1989). However, another recommendation suggested that metal emissions should be 'completely terminated' (Kryuchkov, 1989).

The different technical scenarios leading to these emission levels may result in different types of damage to forest ecosystems. Several models have been used to estimate the optimum pattern of emission reductions, including from all sources uniformly, from low sources alone, from high sources alone, at both industrial sites in Nikel and Zapolyarnyy, or at each of them separately. In addition, the effects on the estimated long-term pollution and on pollution episodes have been estimated (Baklanov et al., 1993; Baklanov and Sivertsen, 1994). The results for Pechenganikel (Baklanov and Sivertsen, 1994) indicate that the critical levels for forests will not be exceeded if SO_2 emissions are reduced by 97% from low sources and 92% from high sources (based on 1990 levels), which means that the Pechenganikel smelter should emit less than 19 kt annually. For the Severonikel smelter (Baklanov et al., 1993), emissions from both low and high sources should be reduced by 95%, i.e. to less than 13 kt of SO_2. At the same time, episodes of high concentrations should be reduced by 98% of the 1990 level.

If the processing of the Noril'sk ores is continued, the smelters will need to be completely reconstructed and high stacks (> 150 m) introduced if nature protection requirements are to be met (Sivertsen and Bekkerstad, 1996). If emissions of the Pechenganikel are reduced to 10 kt SO_2, critical levels will not be exceeded at any distance from the smelter (Sivertsen and Bekkerstad, 1996).

4.5   Possible emission scenarios

Russia joined the Convention on Long-range Transboundary Air Pollution in 1985 (see Chapter 12, this volume, for further details). The Action Programme for Limiting Air Pollution in Areas near the Common Border was signed in 1989 (Bärlund and Katushev, 1990). Both sides committed themselves to reducing total sulphur emissions by 50% from the 1980s levels by the end of 1995. Although sulphur emissions from the Kola smelters had dropped by 37% by 1995 without any investments in environmental protection measures, the emissions may increase in future if there is any increase in metal production.

The modernization of the Pechenganikel smelter is one of the key projects for the Barents Region cooperation (Bröms, 1997). In 1995, Norway and Russia signed an agreement on the reduction of emissions from the Pechenganikel smelter with the financial assistance of Norway. However, following delays with implementation of the project, the Norwegian government has recalled the money (Bröms, 1997). About half the gas-cleaning systems were either not being used or were not working properly in 1995 (Murmansk Regional Committee on Ecology, 1996). Consequently, the proper use of existing facilities could significantly decrease emissions. Recent technology would enable a reduction in emissions by 50% without rebuilding the smelters (Pozniakov, 1993). Another estimate has suggested that emissions from the Pechenganikel smelter under current technologies could be reduced by a factor of 10–12, whereas emissions from the Severonikel could be reduced to one-third of the current level if financial obstacles were removed (Peshev, 1994). However, the current reconstruction plans for the Pechenganikel smelter will only assure a decrease in emissions to 65 kt of SO_2 (Blatov et al., 1995), 68% of the 1995 level. Further declines in emissions are likely to result from future reductions in metal production, which are expected to decrease to 80–85% of the 1992–1993 level (Peshev, 1994).

The Nordic countries, especially Norway and Finland, have been concerned about the smelting industry on Kola since 1978. Several possibilities for the Pechenganikel reconstruction (the technologies of Vanyukov (Russia), Outukumpu (Finland) and the Pechenganikel Reconstruction Consortium (Norway, Sweden and Russia)) associated with 'acceptable' levels of future emissions have been suggested (Dickman, 1990; Bröms, 1997). However, Bröms (1997) has suggested that financial problems within the Russian Federation arising from disputes between central government and the regional authorities have prevented these projects from materializing.

4.6   Tentative predictions

The first predictions on the future of the environment around the Severonikel smelter (Kryuchkov and Syroid, 1984) considered two situations: emissions were kept constant ('business as usual'), and emissions decreased with time. In the first situation, the impact zone was expected to increase by c. 3% annually (Kryuchkov, 1988); the second scenario predicted a decrease in the area of the impact zone and vegetation recovery. It is now clear that these prognoses were too optimistic: although emissions steadily declined between 1983 and 1994 (Murmansk Regional Committee on Ecology, 1996), the impact zone continued to increase until very recently (Pozniakov, 1993; Tsvetkov, 1993; Tikkanen and Niemelä, 1995; Mikkola, 1996). A recent model of forest degradation under emission pressure suggests that even in the unlikely event that emissions are terminated, the spread of the degradation zones around the Severonikel smelter may continue for 10–20 years (Tarko et al., 1995).

Another concern is the time necessary for the natural leaching of metal pollutants from heavily contaminated soils. Assuming that emissions are completely terminated, it may take 90 years before the concentration of nickel returns to background values; for copper, the expected time is 200 years and for cobalt it is 58 years (Evdokimova, 1995). Barkan *et al.* (1993a), however, estimate that metal leaching will take 100–270 years.

5 Conclusion

The copper–nickel smelters in the Kola Peninsula have determined the pattern of contamination in an area extending 100 km from the sources. They emit 350–750 kt of sulphur dioxide annually and have caused visible damage to forest over 39,000 km^2 and forest death over 600–1000 km^2. Individual and population responses of the forest-forming woody plants (Scots pine, Norway spruce and mountain birch) to the multi-component emissions of these smelters are reasonably well documented, whereas changes in other components of forest ecosystems are less known, and virtually no data exist about pollution effects on the functioning of the entire ecosystem. The causal links behind the pollution-driven degradation of the forest ecosystems are poorly understood, and the roles of most of the individual components of emissions in environmental degradation are virtually unknown. The spatial representativeness of study sites used to evaluate effects of emissions on terrestrial communities has been insufficient for the detailed mapping of forest health in the region. The use of the remote sensing analysis supported by ground truth data has overcome this problem and has provided insights into the history of pollution impact, at least during the last 20 years. Retrospective analyses of tree increment, as well as comparisons of the recent state of vegetation with the pre-industrial situation, which was documented for a number of study sites in 1920 to the 1930s, are other tools for understanding the temporal dynamics of forest degradation. Although there are several problems which influence long-term predictions of ecosystem development under different pollution scenarios, forest degradation is likely to continue for some time, even if emissions are terminated.

Acknowledgements

O.R. is grateful to CMF, Sweden, for financial support. M.V.K. acknowledges the financial support from the Academy of Finland (RESTORE program) and from the European Community. This work is a contribution to the EU ELOISE Programme (ELOISE No. 035) in the framework of the BASIS project carried out under contract ENV4-CT97-0637. Special thanks are due to V. Alexeyev, A. Baklanov, E. Zvereva and J. Lesinski for commenting on an earlier version of the manuscript.

References

Aamlid, D., Tømmervik, H., Gytarsky, M., Karaban, R., Venn, K., Rindal, T., Vassilieva, N., Koptsik, G. and Løbersli, E. (1995) Determination of exceedance of critical levels in the border area between Norway and Russia. In: Løbersli, E. and Venn, K. (eds) *Effects of Air Pollutants on Terrestrial Ecosystems in the Border Area Between Russia and Norway. Proceedings of the II Symposium, Svanvik, Norway, 3–5 October 1994*. Statens Forurenssningstilsyn, Trondheim, pp. 19–23.

Ablaeva, Z.C. (1978) Industrial pollution and distribution of lichens in Monchegorsk region. In: Martin, J.L. (ed.) *Lichenoindication of the Environment: Abstracts of the Conference, August 22–27, 1978*. Tallinn Botanical Garden, Tallinn, pp. 129–132 (in Russian).

Alekseev, A.S. (1993) Radial increment of trees and stands under aerial pollution impact. *Lesovedenie [Forestry*, Moscow] 0 (4), 66–70 (in Russian).

Alekseev, A.S. (1995) Mathematical modelling of spatial density of plant damage caused by air pollution. *Ekologija [Ecology*, Sverdlovsk] 0 (6), 428–436 (in Russian).

Alexeyev, V.A. (1989) Forest health diagnostics and its application to studies of air pollution impact. In: Noble, R.D., Martin, J.L. and Jensen, K.F. (eds) *Air Pollution Effects on Vegetation, Including Forest Ecosystems. Proceedings of the Second US-USSR Symposium, September 1989*. US Department of Agriculture, Forest Service, North-eastern Forest Experiment Station, Corvallis, Oregon, pp. 135–141.

Alexeyev, V.A. (1990a) Some problems of diagnostics and classification of injured forest ecosystems. In: Alexeyev, V.A. (ed.) *Forest Ecosystems and Atmospheric Pollution*. Nauka, Leningrad, pp. 38–54 (in Russian).

Alexeyev, V.A. (ed.) (1990b) *Forest Ecosystems and Atmospheric Pollution*. Nauka, Leningrad (in Russian).

Alexeyev, V.A. (1993) Air pollution impact on forest ecosystems of Kola peninsula: history of investigations, progress and shortcomings. In: Kozlov, M.V., Haukioja, E. and Yarmishko, V.T. (eds) *Aerial Pollution in Kola Peninsula. Proceedings of the International Workshop, 14–16 April 1992, St Petersburg, Russia*. Kola Science Centre, Apatity, pp. 20–34.

Alexeyev, V.A. (1995) Impacts of air pollution on far north forest vegetation. *Science of the Total Environment* 160/161, 605–617.

AMAP (1997) *The Arctic Monitoring and Assessment Programme International Symposium on Environmental Pollution in the Arctic, 1–5 June 1997, Tromsø, Norway*. Extended abstracts. Tromsprodukt AS, Tromsø.

Andreeva, E.N. (1989) Changes in moss cover of the northern forests under the impact of industrial pollution. PhD thesis, V. L. Komarov Botanical Institute, Leningrad (in Russian).

Anonymous (1991) *Forests of the U.S.S.R.: Statistical Data (as of 1.01.1988)*. The State Committee of Forestry of the USSR, Moscow (in Russian).

Anonymous (1994) Europe's top 100 polluters. *New Scientist* 143, 9.

Anonymous (1996) Barents Region. *Green Issues, January 1996*. Statens Forurensningstilsyn, Oslo (in Russian).

Armand, A.D., Kaidanova, V.V., Kushnareva, G.V. and Dobrodeev, V.G. (1991) Determination of geosystem stability limits for the surroundings of Monchegorsk metallurgical enterprise. *Izvestija Akademii Nauk SSSR* [*Bulletin of the Academy of Sciences of the USSR*], Geographical Series 0 (1), 93–104 (in Russian).

Baklanov, A. (1994) Problems of airborne pollution at the Kola peninsula. *European Association for the Science of Air Pollution, Newsletter* 23, 5–11.

Baklanov, A.A. and Rodyushkina, I.A. (1993) Pollution of ambient air by 'Severonikel' smelter complex: observations and modelling. In: Kozlov, M.V., Haukioja, E. and Yarmishko, V.T. (eds) *Aerial Pollution in Kola Peninsula. Proceedings of the International Workshop, 14–16 April 1992, St Petersburg, Russia.* Kola Science Centre, Apatity, pp. 83–89.

Baklanov, A.A. and Sivertsen, B. (1994) *Investigations of Local Transborder Air Pollution.* Report F–19/94. NILU, Kjeller.

Baklanov, A., Klyuchnikova, E., Rodyushkina, I. and Smagin, A. (1993) The monitoring and modelling of atmosphere pollution in industrial districts of Kola North. In: Sivertsen, B. (ed.) *Air Pollution Problems in the Northern Region of Fennoscandia Including Kola. Proceedings of the Seminar, 1–3 June 1993, Svanvik, Norway.* NILU, Kjeller, pp. 20–40.

Baklanov, A., Berlin, V., Klyuchnikova, E., Limkina, S., Makarova, T., Rodyushkin, I. and Rodyushkina, I. (1994) Monitoring and modelling of SO_2 and heavy metals in the atmosphere of the Kola peninsula in accordance with Russian–Norwegian programme on co-operation. Report for the year 1993, Apatity, Kola Science Centre. Unpublished manuscript (in Russian).

Baklanov, A., Jaffe, D., Mahura, A., Torvela, H. and Beloglazov, M. (1997) The Monchegorsk Arctic Mesoscale Experiment – measurement and modelling of atmospheric pollution over a complex terrain. *'Air Pollution '97', International Conference, 16–19 September 1997, Bologna, Italy.* FOA Report FOA-B-97-00251-862-SE. FOA, Umeå.

Barkan, V.S., Pankratova, R.P., Silina, A.V. and Koshurnickov, A.B. (1993a) The lysis of nickel and copper from the soil contaminated by metallurgical dust. In: Christie, S.J. and Martin, J. (eds) *Abstracts of Lectures from the International Symposium on the Ecological Effects of Arctic Airborne Contaminants, 4–8 October 1993, Reykjavik, Iceland,* Special Report 93–23. Cold Region Research and Engineering Laboratory, Hanover, New Hampshire, p. 6.

Barkan, V.Sh., Pankratova, R.P. and Silina, A.V. (1993b) Soil contamination by nickel and copper in area polluted by 'Severonikel' smelter complex. In: Kozlov, M.V., Haukioja, E. and Yarmishko, V.T. (eds) *Aerial Pollution in Kola Peninsula: Proceedings of the International Workshop, 14–16 April 1992, St Petersburg.* Kola Science Centre, Apatity, pp. 119–147.

Bärlund, K. and Katushev, K.F. (1990) *The Action Program for Limiting Air Pollution in Areas near the Finnish-Soviet Border,* 26 October 1989, No. 30–35. Statens tryckericentral, Helsinfors, pp. 152–156.

Berlin, V.E. (1993) Analysis of ambient air at weather station 'Chunozero' in Lapland biosphere reserve. In: Kozlov, M.V., Haukioja, E. and Yarmishko, V.T. (eds) *Aerial Pollution in Kola Peninsula. Proceedings of the International Workshop, 14–16 April 1992, St Petersburg, Russia.* Kola Science Centre, Apatity, pp. 99–104.

Blatov, I.A., Ezhov, E.I., Nosan, L.M., Mushkatin, L.M., Tsemekhman, L.S., Klementjev, V.V. and Mosiondz, K.I. (1995) Strategies of the increase in efficiency and solution of ecological problems of the 'Pechenganikel' smelter on the basis of reconstruction of the metallurgical process. *Tsvetnye Metally* [*Non-Ferrous Metals*, Moscow] 0 (8), 20–22 (in Russian).

Bobrinsky, A. (1993) Mapping of forest conditions in Murmansk region. In: Kozlov, M.V., Haukioja, E. and Yarmishko, V.T. (eds) *Aerial Pollution in Kola Peninsula. Proceedings of the International Workshop, 14–16 April 1992, St Petersburg, Russia.* Kola Science Centre, Apatity, pp. 407–408.

Brodin, Y.-W. and Kuylenstierna, J. (1992) Acidification and critical loads in Nordic countries: a background. *Ambio* 21, 332–338.

Bröms, P. (1997) *Crossing the Thresholds: The Forming of an Environmental Security Regime in the Arctic North.* Research Report 28 1997, The Swedish Institute of International Affairs, Stockholm.

Buznikov, A.A., Payanskaya-Gvozdeva, I.I., Jurkovskaya, T.K. and Andreeva, E.N. (1995a) Use of remote and ground methods to assess the impacts of smelter emissions in the Kola peninsula. *Science of the Total Environment* 160/161, 285–293.

Buznikov, A.A., Lakhtanov, G.A., Alexeyeva-Popova, N.V., Payanskaya-Gvozdeva, I.I., Virolainen, A.V. and Shvareva, S.G. (1995b) Investigation of reflecting spectra of the indicator mosses. *Issledovanie Zemli iz Kosmosa* [*Investigation of the Earth from Space*, Moscow] 0 (2), 37–45 (in Russian).

CAFF (1996) *Conservation of Arctic Flora and Fauna. Proposed Protected Areas in the Circumpolar Arctic 1996.* CAFF Conservation Report No 2, Directorate for Nature Management, Trondheim, Norway.

Chekrizov, E.A. (1987) Changes in the structure of on-ground combustibles in young pine stands affected by industrial pollutions in Murmansk region. In: *Materials of the Session Reporting Results of Research Conducted by the Arkhangelsk Institute of Forestry and Forest Chemistry in 1986.* Arkhangelsk Institute of Forestry and Forest Chemistry, Arkhangelsk, pp. 81–82 (in Russian).

Dickman, S. (1990) Cleaning Kola with kroner. *Nature* 346, 787.

Doiban, V.A., Pretes, M. and Sekarev, A.V. (1992) Economic development in the Kola region, USSR: an overview. *Polar Record* 28, 7–16.

Doncheva, A.V. (1978) *The Landscape in Zone of Industrial Impact.* Lesnaya Promyshlennost, Moscow (in Russian).

Doncheva, A.V., Kazakova, L.K. and Kalutskov, V.N. (1992) *The Landscape Indications of Environmental Contamination.* Ekologija, Moscow (in Russian).

Downing, R., Hettelingh, J.-P. and de Smet, P. (eds) (1993) *Convention on Long-range Transboundary Air Pollution. Calculation and Mapping of Critical Loads in Europe. Status Report.* Coordination Centre for Effects, National Institute of Public Health and Environmental Protection, Bilthoven, The Netherlands.

Evdokimova, G.A. (1993) Anthropogenic impacts on soil. In: Kalabin, G.V. and Evdokimova, G.A. (eds) *Ecology and Nature Protection of the Kola North.* Kola Science Centre, Apatity, pp. 123–131 (in Russian).

Evdokimova, G.A. (1995) *Ecological and Microbiological Foundations of Soil Protection in the Far North.* Kola Science Centre, Apatity (in Russian).

Evseev, A.V. and Krasovskaya, T.M. (1990) Zonation of the Kola peninsula for the organization of the environmental monitoring. In: Evdokimov, S.P. (ed.) *Information Aspects of Regional Use of Natural Resources.* Moscow State University, Saransk, pp. 83–88 (in Russian).

Fedorkov, A. (1992) Genetic monitoring of Scots pine stands on the Kola peninsula. In: Tikkanen, E., Varmola, M. and Katermaa, T. (eds) *Symposium on the State of the*

Environment and Environmental Monitoring in Northern Fennoscandia and the Kola Peninsula. 6–8 October 1992, Rovaniemi, Finland, Extended Abstracts. Arctic Centre Publications Vol. 4, Arctic Centre, Rovaniemi, pp. 256–258.

Georgievskii, A.B. (1990) The influence of aerial pollution on vegetation of suburban area of Kandahlaksha and Kandalaksha reserve. *Biologicheskie Nauki [Biological Sciences]* 0 (9), 124–133 (in Russian).

Gorshkov, V.V. (1993) Epiphytic lichens in polluted and unpolluted pine forests of the Kola peninsula. In: Kozlov, M.V., Haukioja, E. and Yarmishko, V.T. (eds) *Aerial Pollution in Kola Peninsula. Proceedings of the International Workshop, 14–16 April 1992, St Petersburg, Russia.* Kola Science Centre, Apatity, pp. 290–298.

Gurevich, V.I. (1966) Hydro-chemical investigations of aero-technogenic damage in the Monchegorsk district. Kola Science Centre, Apatity (unpublished manuscript, in Russian).

Hagner, O. and Rigina, O. (1998) Detection of forest decline in Monchegorsk area. *Remote Sensing of the Environment* 62, 11–23.

Høgda, K.A. and Tømmervik, H. (1996) Mapping of air pollution effects on the vegetation cover in the Kirkenes–Nikel area applying satellite remote sensing. *Pollution Monitoring* 4, 18–19.

Høgda, K.A., Tømmervik, H., Solheim, I. and Marhaug, Ø. (1995) *Use of Multitemporal Landsat Image Data for Mapping the Effects From Air Pollution in the Kirkenes-Pechenga Area in the Period 1973–1994.* NORUT Informasjonsteknologi, Tromsø.

Jaffe, D., Cerundolo, B., Rickers, J., Stolzberg, R. and Baklanov, A. (1995) Deposition of sulphate and heavy metals on the Kola peninsula. *Science of the Total Environment* 160/161, 127–134.

Johansen, M.E., Tømmervik, H., Guneriussen, T. and Pedersen, J.P. (1995) Using a Geographic Information System (ARC/INFO) to integrate remote sensed and in-situ data in an analysis of the air pollution effects on terrestrial ecosystems in the border areas between Norway and Russia. In: Løbersli, E. and Venn, K. (eds) *Effects of Air Pollutants on Terrestrial Ecosystems in the Border Area Between Russia and Norway. Proceedings of the II Symposium, 3–5 October 1994, Svanvik, Norway.* Statens Forurensningstilsyn, Trondheim, pp. 120–130.

Kauhanen, H. and Varmola, M. (eds) (1992) Itä-Lapin metsävaurioprojektin välira-portti. The Lapland Forest Damage Project. Interim Report. *Metsäntutkimus-laitoksen Julkaisuja* 413, 1–266.

Kismul, V., Jerre, J. and Løbersli, E. (eds) (1992) *Effects of Air Pollutants on Terrestrial Ecosystems in the Border Area Between Russia and Norway. Proceedings of the I Symposium, 18–20 March 1992, Svanvik, Norway.* Statens Forurensningstilsyn, Oslo.

Koptsik, G. and Koptsik, S. (1995) Assessment of critical loads of acid deposition for soils in Kola peninsula. In: Løbersli, E. and Venn, K. (eds) *Effects of Air Pollutants on Terrestrial Ecosystems in the Border Area between Russia and Norway. Proceedings of the II Symposium, 3–5 October 1994, Svanvik, Norway,* Report 1995–8, Directorate for Nature Management, Trondheim, pp. 54–60.

Koroleva, N. (1993) Pollution-induced changes in forest vegetation structure as revealed by ordination test. In: Kozlov, M.V., Haukioja, E. and Yarmishko, V.T. (eds) *Aerial Pollution in Kola Peninsula. Proceedings of the International Workshop, 14–16 April 1992, St Petersburg, Russia.* Kola Science Centre, Apatity, pp. 339–345.

Kozlov, M.V. and Haukioja, E. (1995) Pollution-related environmental gradients around the 'Severonikel' smelter complex at the Kola peninsula, Northwestern Russia. In: Munawar, M. and Luotola, M. (eds) *The Contaminants in the Nordic Ecosystem: Dynamics, Processes and Fate*. Ecovision World Monograph Series, SPB Academic Publishing, Amsterdam, pp. 59–69.

Kozlov, M.V. and Haukioja, E. (1997) Microclimate changes along a strong pollution gradient in northern boreal forest zone. In: Uso, J.L., Brebbia, C.A. and Power, H. (eds) *Ecosystems and Sustainable Development*. Advances in Ecological Sciences, Vol. 1. Computation Mechanics Publishing, Southampton, pp. 603–614.

Kozlov, M.V. and Niemalä, P. (1999) Needle fluctuating asymmetry as an objective indicator of pollution impact on Scots pine (*Pinus sylvestris*). *Water, Air, and Soil Pollution* (in press).

Kozlov, M.V., Haukioja, E. and Yarmishko, V.T. (eds) (1993) *Aerial Pollution in Kola Peninsula. Proceedings of the International Workshop, 14–16 April 1992, St Petersburg, Russia*. Kola Science Centre, Apatity.

Kozlov, M.V., Wilsey, B.J., Koricheva, J. and Haukioja, E. (1996) Fluctuating asymmetry of birch leaves increases under pollution impact. *Journal of Applied Ecology* 33, 1489–1495.

Kozlov, M.V., Zvereva, E.L. and Niemalä, P. (1999) Effects of soil quality and air pollution on rooting and survival of *Salix borealis* cuttings. *Boreal Environment Research* 4, 67–76.

Kravtsova, V.I., Lurje, I.K. and Ressle, A.I. (1995) Remote sensing of the consequences of industrial impacts on vegetation on the basis of multi-spectral satellite data (an example from the Monchegorsk area). *Geografija i Prirodnye Resursy* [*Geography and Natural Resources*] 0 (3), 149–158 (in Russian).

Kryuchkov, V.V. (1988) We also need perestroika in nature protection. *Sever* [*North*, Moscow] 0 (1), 85–89 (in Russian).

Kryuchkov, V.V. (1989) Real and permissible pollution loads on ecosystems of the Kola North. In: Anon. (ed.) *Problems with the All-Round Utilization of the Natural Resources of the Kola Peninsula. Abstracts of Lectures from the All-Union Conference in Apatity, 30 November – 2 December 1989*. Kola Science Centre, Apatity, pp. 98–99 (in Russian).

Kryuchkov, V.V. (1993) Extreme anthropogenic loads and the northern ecosystem condition. *Ecological Applications* 3, 622–630.

Kryuchkov, V.V. and Makarova, T.A. (1989) *Impact of Industrial Air Pollution on Ecosystems of the Kola North*. Kola Science Centre, Apatity (in Russian).

Kryuchkov, V.V. and Syroid, N.A. (1984) Soil and botanical monitoring in the central part of the Kola Region. In: Kryuchkov, V.V. (ed.) *Monitoring of the Kola North Environment*. Kola Science Centre, Apatity, pp. 15–26 (in Russian).

Kryuchkov, V.V. and Syroid, N.A. (1987) Transformation of boreal-taiga ecosystems in industrial districts of the North. In: Vorontsov, A.I. (ed.) *Protection of Forest Ecosystems and Rational Utilisation of Forest Resource, Extended Abstracts of the Conference, 20–22 October 1987, Forest Technological Institute, Moscow, Russia*, pp. 13–14 (in Russian).

Lange, M.A. (1996) *Assessing the Consequences of Global Changes for the Barents Region: the Barents Sea Impact Study (BASIS)*. University of Lapland, Rovaniemi.

Lukina, N.V. and Nikonov, V.V. (1993) *State of the Northern Spruce Biogeocenoses Exposed to Industrial Pollution*. Kola Science Centre, Apatity (in Russian).

Lukina, N.V. and Nikonov, V.V. (1996) *Biogeochemical Cycles in the Northern Forests Subjected to Air Pollution*, Parts 1 and 2. Kola Science Centre, Apatity (in Russian).

Manninen, S. and Huttunen, S. (1997) Critical level of SO_2 for subarctic Scots pine forests. In: Anon. (ed.) *The AMAP International Symposium on Environmental Pollution in the Arctic. 1–5 June 1997, Tromsø, Norway*. Extended Abstracts, Tromsprodukt AS, Tromsø, pp. 173–174.

Mikkola, K. (1996) A remote sensing analysis of vegetation damage around metal smelters in the Kola peninsula, Russia. *International Journal of Remote Sensing* 17, 3675–3690.

Mikkola, K. and Ritari, A. (1992) Satellite survey of forest damage in the Monchegorsk area, Kola peninsula. In: Tikkanen, E., Varmola, M. and Katermaa, T. (eds) *Symposium on the State of the Environment and Environmental Monitoring in Northern Fennoscandia and the Kola Peninsula. 6–8 October 1992, Rovaniemi, Finland*, Extended Abstracts, Arctic Centre Publications Vol. 4. Arctic Centre, Rovaniemi, pp. 310–313.

Moiseenko, T. (1994) Acidification and critical loads in surface waters: Kola, Northern Russia. *Ambio* 23, 418–424.

Mortensen, L.M. (1993) Effects of ozone on young plants of *Betula pubescens* Ehrh. at different temperature and light levels. *Norwegian Journal of Agricultural Sciences* 7, 293–300.

Mortensen, L.M. and Nilsen, J. (1992) Effects of ozone and temperature on growth of several wild plant species. *Norwegian Journal of Agricultural Sciences* 6, 195–204.

Murmansk Regional Committee on Ecology (1996) *The State and Protection of the Environment in the Murmansk Region in 1995*. Ministry of Environmental Protection and Natural Resources of Russia, Murmansk (in Russian).

Murmansk Regional State Committee on Nature Protection (1997) *State of Nature and Ecological Problems in the Kola Peninsula in 1996*. Ministry of Environmental Protection and Natural Resources of Russia, Murmansk (in Russian).

Neshataev, V.Yu. and Neshataeva, V.Yu. (1993) Forest vegetation of Ponoi river valley (the unpolluted area). In: Kozlov, M.V., Haukioja, E. and Yarmishko, V.T. (eds) *Aerial Pollution in Kola Peninsula: Proceedings of the International Workshop, 14–16 April 1992, St Petersburg*. Kola Science Centre, Apatity, pp. 346–360.

Neuvonen, S. (1993) Experiences from a field experiment of the effects of simulated acid rain in Finnish Lapland. In: Kozlov, M.V., Haukioja, E. and Yarmishko, V.T. (eds) *Aerial Pollution in Kola Peninsula. Proceedings of the International Workshop, 14–16 April 1992, St Petersburg*. Kola Science Centre, Apatity, pp. 70–73.

Nikonov, V.V. and Lukina, N.A. (1994) *Biogeochemical Functions of Forests on the Northern Tree Line*. Kola Science Centre, Apatity (in Russian).

Nilsson, A. (1997) *AMAP. Arctic Pollution Issue: A State of the Arctic Environmental Report*. AMAP, Oslo.

Nilsson, S. and Duinker, P. (eds) (1988) *Seminar on Remote Sensing of Forest Decline Attributed to Air Pollution, 11–12 March 1987, Laxenburg, Austria: Proceedings*. International Institute for Applied Systems Analysis, Laxenburg.

Norby, R.J. and Kozlowski, T.T. (1981) Interaction of SO_2 concentration and post-fumigation temperature on growth of woody plants. *Environmental Pollution* 25, 27–39.

Norin, B.N. and Yarmishko, V.T. (eds) (1990) *The Influence of Industrial Atmospheric Pollution on Pine Forests of the Kola Peninsula.* Komarov Botanical Institute, Leningrad (in Russian).

Pastukhova, E., Golovina, M. and Karpova, N. (1992) Estimation of terrestrial ecosystems stability (sensitivity) of Kola peninsula to the influence of acidifying deposition. In: Kismul, V., Jerre, J. and Løbersli, E. (eds) *Effects of Air Pollutants on Terrestrial Ecosystems in the Border Area Between Russia and Norway. Proceedings of the II Symposium, 18–20 March 1992, Svanvik, Norway.* Statens Forurensningstilsyn, Oslo, pp. 145–153.

Perkins, T.D., Vogelmann, H.W. and Klein, R.M. (1987) Changes in light intensity and soil temperature as a result of forest decline on Camels Hump, Vermont. *Canadian Journal of Forest Research* 17, 565–568.

Peshev, N.G. (1994) The northern nickel of Russia. In: Peshev, N.G. (ed.) *Economic Evaluation of Natural Resource Use Under the Conditions of a Market Economy.* Kola Science Centre, Apatity, pp. 39–53 (in Russian).

Plakunova, V.G., Plakunova, O.V. and Gusev, M.V. (1983) Physiology of epigeic lichens in relation to the early indication of the environment pollution. *Izvestiya AN SSSR [Bulletin of the Academy of Science of the USSR], Biological Series* 0 (6), 888–896 (in Russian).

Pozniakov, V.Ya. (1993) The 'Severonikel' smelter complex: history of development. In: Kozlov, M.V., Haukioja, E. and Yarmishko, V.T. (eds) *Aerial Pollution in Kola Peninsula. Proceedings of the International Workshop, 14–16 April 1992, St Petersburg, Russia.* Kola Science Centre, Apatity, pp. 16–19.

Raitio, H. (1992) The foliar chemical composition of Scots pines in Finnish Lapland and on the Kola peninsula. In: Tikkanen, E., Varmola, M. and Katermaa, T. (eds) *Symposium on the State of Environment and Environmental Monitoring in the Northern Fennoscandia and the Kola Peninsula, 6–8 October 1992, Rovaniemi, Finland,* Extended Abstracts, Arctic Centre Publications Vol. 4. Arctic Centre, Rovaniemi, pp. 226–231.

Ratkin, N.E. (1993) Pollution of air basin of the Murmansk region. Vegetation state. In: Kalabin, G.V. and Evdokimova, G.A. (eds) *Ecology and Nature Protection of the Kola North.* Kola Science Centre, Apatity, pp. 111–122 (in Russian).

Rees, W. and Williams, M. (1997) Monitoring changes in land cover induced by atmospheric pollution in the Kola peninsula, Russia, using Landsat-MSS data. *International Journal of Remote Sensing* 18, 1703–1723.

Reimann, C., Chekushin, V.A. and Äyräs, M. (eds) (1996) *Kola Project – International Report, Catchment study 1994,* Report 96.088. Geological Survey of Norway, Trondheim.

Rigina, O. and Baklanov, A. (1998) Trends in sulphur emission induced effects in Northern Europe. *Water, Air and Soil Pollution* 15, 331–342.

Rigina, O., Baklanov, A., Hagner, O. and Olsson, H. (1999) Monitoring of forest damage in the Kola Peninsula, northern Russia, due to smelting industry. *Science of the Total Environment* 229, 147–163.

Roberts, D., Karputs, R., Tømmervik, H., Johansen, B.E. and Pedersen, J.P. (1994) Spectral detection of mining-induced environmental stress in the Nikel-Zapolyarny region of western Kola peninsula, NW Russia, and adjacent areas of Norway. In: *Proceedings of the 10th Thematic Conference on Geologic Remote Sensing, 9–12 May 1994, San Antonio, Texas, USA,* Vol. 2, pp. 109–120.

Ruohomäki, K., Kaitaniemi, P., Kozlov, M.V., Tammaru, T. and Haukioja, E. (1996) Density and performance of *Epirrita autumnata* (Lep., *Geometridae*) along three air pollution gradients in northern Europe. *Journal of Applied Ecology* 33, 773–785.

Sakai, A. and Otsuka, K. (1970) Freezing resistance of alpine plants. *Ecology* 51, 665–671.

Scherbatyuk, L.K. (1988) Effects of atmosphere pollution with sulphur dioxide on forest ecosystems. *Proceedings of the State Nikitskij Botanical Garden, Yalta*, pp. 73–83 (in Russian).

Sharapa, T.V. and Pisareva, S.D. (1987) Species composition of tree-eating insects in forests of the Murmansk region affected by industrial pollution. In: Vorontsov, A.I., Teodoronsky, V.S. and Nikolaevskaja, N.G. (eds) *Problems of Forest Protection and Greening of Cities. Nauchnye trudy Moskovskogo Lesotekhnicheskogo Instituta* [*Proceedings of the Moscow Institute of Forest Technologies*] 188, 56–60 (in Russian).

Sivertsen, B. and Bekkerstad, T. (1996) *Konsekvenser for Luftkvaliteten ved Utslipp fra Ombygd Smelteverket i Nikel. Alternative Losninger*, Report O–1845,4/96. NILU, Kjeller.

Sivertsen, B., Baklanov, A., Hagen, L. and Makarova, T. (1994) *Air Pollution in the Border Areas of Norway and Russia. Summary Report 1991–1993*, Report OR 56/94. NILU, Kjeller.

Sokolovsky, V.G. (ed.) (1990) *The State of the Environment in the U.S.S.R. in 1988: a Report*. Lesnaya promyshlennost, Moscow (in Russian).

Solheim, I., Tømmervik, H. and Høgda, K. (1995) *A Time Study of the Vegetation in Monchegorsk, Russia*. Report IT2039/2-95. NORUT–IT, Tromsø.

Stavrova, N.I. (1990) Renewal of Scots pine (*Pinus sylvestris* L.) under the impact of industrial pollution in north taiga forests. PhD thesis, V. L. Komarov Botanical Institute, St Petersburg (in Russian).

Sutinen, S. and Koivisto, L. (1992) Cell injuries in the needles of Scots pine (*Pinus sylvestris* L.) in Finnish Lapland. A light and electron microscopic approach. In: Tikkanen, E., Varmola, M. and Katermaa, T. (eds) *Symposium on the State of the Environment and Environmental Monitoring in the Northern Fennoscandia and the Kola Peninsula, 6–8 October 1992, Rovaniemi, Finland*, Extended Abstracts, Arctic Centre Publications Vol. 4. Arctic Centre, Rovaniemi, pp. 232–234.

Sutinen, M.-L., Raitio, H., Nivala, V., Ollikainen, R. and Ritari, A. (1996) Effects of emissions from copper-nickel smelters on the frost hardiness of *Pinus sylvestris* needles in the subarctic region. *New Phytologist* 132, 503–512.

Tarhanen, S., Holopainen, T., Poikolainen, J. and Oksanen, J. (1996) Effect of industrial emissions on membrane permeability of epiphytic lichens in northern Finland and the Kola peninsula industrial areas. *Water, Air, and Soil Pollution* 88, 189–201.

Tarko, A.M., Bykadorov, A.V. and Kryuchkov, V.V. (1995) Modelling of air pollution impact on forest ecosystems in region. *Doklady RAN* [*Doklady of the Russian Academy of Science, Moscow*] 341, 571–573 (in Russian).

Tikkanen, E. and Niemelä, I. (eds) (1995) *Kola Peninsula Pollutants and Forest Ecosystems in Lapland. Final Report of the Lapland Forest Damage Project*. Finnish Forest Research Institute, Rovaniemi.

Tikkanen, E., Varmola, M. and Katermaa, T. (eds) (1992) *Symposium on the State of the Environment and Environmental Monitoring in Northern Fennoscandia and the*

Kola Peninsula, 6–8 October 1992, Rovaniemi, Finland, Extended Abstracts, Arctic Centre Publications Vol. 4. Arctic Centre, Rovaniemi.

Tømmervik, H., Johansen, B.E. and Pedersen, J.P. (1995) Monitoring the effects of air pollution on terrestrial ecosystems in Varanger (Norway) and Nikel–Pechenga (Russia) using remote sensing. *Science of the Total Environment* 160/161, 753–767.

Tsvetkov, V.F. (1990a) Forest damage caused by emission of non-ferrous smelter in Murmansk region. In: Lukjanova, L.M. (ed.) *Botanical Investigations to the North of the Polar Circle.* Kola Science Centre, Apatity, pp. 185–196 (in Russian).

Tsvetkov, V.F. (1990b) Forests of the Kola peninsula, their state and use. In: Anon. (ed.) *Northern Forests: State, Dynamics, Human Impact. Abstracts of the International Symposium, 16–26 July 1990, Arkhangelsk, Russia.* All Union Scientific Research Centre on Forest Resources of the USSR, Moscow, pp. 90–91 (in Russian).

Tsvetkov, V.F. (1991) State of forests subjected to industrial emission impact in Murmansk region and problems of their preservation. In: *Ecological Research in the Forests of the European North.* Arkhangelsk Institute of Forestry and Forest Chemistry, Arkhangelsk, pp. 125–136 (in Russian).

Tsvetkov, V.F. (1993) Threshold levels of air pollution and forest decline in surroundings of 'Severonikel' smelter complex. In: Kozlov, M.V., Haukioja, E. and Yarmishko, V.T. (eds) *Aerial Pollution in Kola Peninsula. Proceedings of the International Workshop, 14–16 April 1992, St Petersburg, Russia.* Kola Science Centre, Apatity, pp. 397–401.

Tuovinen, J.-P. and Laurila, T. (1993) Variability of sulphur dioxide and ozone concentrations in Northern Finland. In: Sivertsen, B. (ed.) *Air Pollution Problems in the Northern Region of Fennoscandia Including Kola. Proceedings of the Seminar, 1–3 June 1993, Svanvik, Norway.* NILU, Kjeller, pp. 47–56.

Tuovinen, J.-P., Laurila, T., Lattila, H., Ryaboshapko, A. and Korolev, S. (1993) Impact of the sulphur dioxide sources in the Kola peninsula on air quality in Northermost Europe. *Atmospheric Environment* 27A, 1379–1395.

UN-ECE (1988) *ECE Critical Levels Workshop. Bad Harzburg Workshop. Final Report.* Umweltbundesamt, Berlin.

UN-ECE (1993) *The 1993 Report of the Task Force on Mapping,* Report EB.AIR/WG.1/R.85. United Nations Economic Commission for Europe, Geneva.

Vasilenko, V., Nazarov, I., Fridman, S. and Belikova, T. (1992) On the ecological capacity of terrestrial natural ecosystems with respect to pollution deposition. In: Kismul, V., Jerre, J. and Løbersli, E. (eds) *Effects of Air Pollutants on Terrestrial Ecosystems in the Border Area Between Russia and Norway. Proceedings of the I Symposium, 18–20 March 1992, Svanvik, Norway.* Statens Forurensningstilsyn, Oslo, pp. 20–27.

Vassilieva, N.P. (1993) Response of forest ecosystem to aerial emission of 'Pechenganikel' smelter complex. In: Kozlov, M.V., Haukioja, E. and Yarmishko, V.T. (eds) *Aerial Pollution in Kola Peninsula. Proceedings of the International Workshop, 14–16 April 1992, St Petersburg, Russia.* Kola Science Centre, Apatity, pp. 402–406.

Vatkovskij, O.S. (1990) Canopy structure in primary and secondary forests of the Kola peninsula. *Nauchnye trudy Moskovskogo lesotekhnicheskogo instituta* [*Proceedings of the Moscow Institute of Forest Technologies*] 223, 61–65 (in Russian).

Yarmishko, V.T. (1993) The analysis of morphological structure and crown state of *Pinus sylvestris* L. under different levels of air pollution in the Kola peninsula.

In: Kozlov, M.V., Haukioja, E. and Yarmishko, V.T. (eds) *Aerial Pollution in Kola Peninsula. Proceedings of the International Workshop, 14–16 April 1992, St Petersburg, Russia*. Kola Science Centre, Apatity, pp. 236–251.

Yarmishko, V.T. (1997) *Scots Pine and Aerial Pollution in the European North*. V. L. Komarov Botanical Institute, St Petersburg (in Russian).

Zvereva, E.L., Kozlov, M.V. and Haukioja, E. (1997a) Stress responses of *Salix borealis* to pollution and defoliation. *Journal of Applied Ecology* 34, 1387–1396.

Zvereva, E.L., Kozlov, M.V. and Haukioja, E. (1997b) Population dynamics of a herbivore in an industrially modified landscape: case study with *Melasoma lapponica* (Coleoptera: Chrysomelidae). *Acta Phytopathologica et Entomologica Hungarica* 32, 251–258.

The Impact of Heavy Metals on the Microbial Diversity of Podzolic Soils in the Kola Peninsula

5

G.A. Evdokimova

The Institute of the North Industrial Ecology Problems, Kola Science Centre, Russian Academy of Science, Apatity, Russian Federation

The diversity of frequently occurring genera and species of microorganisms in soils polluted by heavy metals (Cu, Ni, Co) and sulphur compounds has been investigated in the Kola Peninsula. The decrease of species diversity of fungi, algae and especially the prokaryotic component of microbial communities, namely the non-spore-forming gram-negative bacteria, cyanobacteria and streptomycetes, has been established. The homeostasis of the microbial communities of podzolic soils has been changed as a result of heavy metal concentrations in the soils amounting to 300–400 mg kg^{-1} for copper and 600–700 mg kg^{-1} for nickel.

1   Introduction

The main function of many microorganisms is the decomposition of organic matter. As a result of the decomposition processes carried out by microorganisms, carbon dioxide necessary for higher plants is constantly returned to the atmosphere. There is a great diversity of species and functions amongst microorganisms and they are the most important gene pool in forest ecosystems. Ecosystems with richer biodiversity have higher resistance to stress (Lockwood and Pimm, 1994). In such ecosystems, species that have been lost may be replaced by others carrying out similar functions.

The diversity of the microbial communities in soils can be measured in a number of ways: as the classic species number index (e.g. Shannon index), by metabolic capabilities, such as through the study of the phospholipid fatty acid

©CAB *International* 2000. *Forest Dynamics in Heavily Polluted Regions*
(eds J.L. Innes and J. Oleksyn)

composition (Baath *et al.*, 1992; Pennanen *et al.*, 1996) or by the genetic diversity index (Torsvik *et al.*, 1990). With prudent use, all diversity measurements provide a way to develop an understanding of the ecology of microorganisms (Atlas, 1984). However, the quantitative assessment of the diversity of microorganisms at the species level is extremely difficult because of the problems with the determination of bacteria to species level and the general vagueness of the species concept in relation to microorganisms (O'Donnell *et al.*, 1994). The microbial communities of different ecosystems can be compared using an index which characterizes the structure of the community – the index of spatial frequency of the occurrence of a species (or genus), which is the ratio of the number of samples in which the species (genus) was identified to the total number of investigated samples (Tresner *et al.*, 1954). I have adopted this index below. The adverse impacts of pollutants on the microbiota are examined by undertaking parallel investigations of non-polluted background soils.

The levels of pollution in the Kola Peninsula are given elsewhere (Rigina and Kozlov, this volume). All the pollutants are deposited on the vegetation, soil and surface water. Soil is less dynamic and has a better developed buffering system than air or water. The soil buffering capacity is interpreted here as its stability in relation to changes in the environment. The effects of pollution in soils may not become apparent for some time because of the large surface areas available for adsorption. The pollution of soil by toxic elements is particularly insidious as it is less evident than erosion by wind and water. In soil, pollutants are transformed from a soluble state into less soluble forms and may become less available to plants. They are incorporated into the soil-forming process through the soil absorption complex. However, the capacity of the soil is not unlimited and a point is reached after which toxic compounds begin to have an adverse effect on plants, animals and humans.

There are a number of characteristics of northern forest soils that influence their responses to pollution by industrial pollutants in the Kola Peninsula:

*   the intensity of energy- and mass-exchange in northern ecosystems tends to be weak, as a result of the low temperatures and the low self-cleaning ability of the natural media;
*   the soil biota has a relatively low diversity and there are therefore limited possibilities for the biological utilization and transformation of pollutants;
*   the soil solutions of the podzols and peats that dominate the soils of the Kola Peninsula are predominantly acidic, increasing the mobility of pollutants and their toxicity;
*   the water regime of the soils contributes to the leaching of elements through the soil profile and increases the risk of groundwater pollution.

In this study, I report on changes in the microbial community structures around the Severonikel smelter and evaluate the dominant diversity of the microorganisms under conditions of soil pollution by heavy metals and sulphur compounds.

2   Materials and Methods

The sample plots were established at various distances from the Severonikel enterprise, which was founded in 1935. The plots were close to the centre of the pollution (0.2 km from the source), in the impact zone (5 km from the source), in the buffer zone (15 km) and in the background zone (50 km). The soil was an Al–Fe–humus podzol. Prior to the establishment of the smelter, the predominant vegetation was northern taiga forest. However, 60 years of industrial activities have resulted in major transformations of the ecosystems in impacted areas. The vegetation and soil characteristics in the sampling areas can be described as follows.

Epicentre. The soil close to the pollution source is structureless and has a bluish-grey metallic feel. The copper content in the upper 0–10 cm varies from 3000 to 6000 mg kg^{-1}, nickel from 7000 to 15,000 mg kg^{-1}, and cobalt from 200 to 800 mg kg^{-1}. Sulphate ion contents range from 600 to 2000 mg kg^{-1}, and the pH$_{H_2O}$ from 5.6 to 6.0.

Impact zone. The woody vegetation is completely destroyed, with only depressed regeneration of birch (*Betula pubescens* ssp. *czerepanovi* (Orlova) Hämet-Ahti). The soil is eroded. Ground vegetation covers about 20% of the surface, and consists mainly of crowberry (*Empetrum hermaphroditum* Hagerup) and wavy hair-grass (*Deschampsia flexuosa* (L.) Schur.). Mosses and lichens are absent. The concentration of copper in the organic horizon ranges from 600 to 1000 mg kg^{-1}, nickel from 1700 to 2600 mg kg^{-1}, cobalt from 300 to 350 mg kg^{-1}, sulphate ion from 400 to 600 mg kg^{-1}; and pH$_{H_2O}$ from 4.8 to 4.9.

Buffer zone. The vegetation consists of pine forest of dwarf-shrub type, with the addition of birch and a predominance of cowberry (*Vaccinium vitis-idaea* L.), bilberry (*Vaccinium myrtillus* L.) and bearberry (*Arctostaphylos uva-ursi* (L.) Spreng.) in the ground cover. Lichens are represented by *Cladonia rangiferina* (L.) Web. and *Cladina mitis* (Sandst.) Hale et W.Culb. The copper content in the organic horizon ranges from 200 to 400 mg kg^{-1}, nickel from 500 to 1000 mg kg^{-1}, cobalt from 40 to 52 mg kg^{-1}, sulphate ion from 150 to 200 mg kg^{-1}, and pH$_{H_2O}$ from 4.5 to 5.0.

Background zone. The vegetation is dominated by dwarf-shrub type pine forest, with the addition of birch. Crowberrry and bearberry dominate among the dwarf shrubs, and the lichen flora contains representatives of the genera *Cladonia*, *Cladina* and *Cetraria*. *Hylocomium* and *Pleurozium* are among the moss genera present. The copper content of the organic horizon is on average 40 mg kg^{-1}, nickel is 80 mg kg^{-1}, cobalt is 30 mg kg^{-1}, sulphate ion is 8 mg kg^{-1}, and the pH$_{H_2O}$ is 5.0–5.6.

The soil samples for the microorganism diversity study were collected as follows. A soil profile was opened up and samples weighing up to 0.5 kg were collected under sterile conditions from the organic horizon (O_L + O_{F+H}). The thickness of the organic horizon was 0–4 cm in the impact zone, 0–13 cm in the buffer zone and 0–7 cm in the background zone. The organic horizon in the pollution epicentre had been destroyed, so samples were taken from the upper layer of the soil at 0–10 cm depth. The sampling was repeated three times in each August between 1990 and 1996. The samples were brought to the laboratory the same day that they were sampled and were kept in the refrigerator. The following day, they were analysed using the plate-counting technique on the surface of nutrient media. Saprophytic non-spore-forming bacteria were cultivated on meat–pepton agar, and bacilli on meat–pepton agar with wort after warming up to 80°C for 10 min. The fungi were cultivated on the wort with streptomycin added to suppress the bacterial flora. Mineral nitrogen forms of bacteria were cultivated on starch–ammonia agar; cellulose-destroying microorganisms on Getchison medium. The quantity and diversity of microorganisms on a low nutrient medium combined with a soil extract was also examined. The agar of the soils corresponds to the habitat of the microorganisms, enabling autochthonous microorganisms to be taken into account. The isolation and determination of microorganisms belonging to various nitrogen-cycle physiological groups was undertaken using liquid-selective nutrient media: a Hiltay medium for the denitrification bacteria, Vinogradsky media for the nitrification bacteria, and the non-nitrogen medium of Vinogradsky for the nitrogen-fixing bacteria of the genera *Clostridium*. Microorganisms with a spatial occurrence frequency in excess of 30% were isolated from the nutrient media. The index of spatial frequency of the occurrence of a species (or genus) was calculated using the following ratio:

$$\text{presence frequency, \%} = \frac{\text{no. of samples in which the species (genus) was found}}{\text{total no. of samples analysed}} \times 100$$

Holt and Zavarzin (1980) were followed for the identification of bacteria, based on phenotypic, physiological and biochemical characteristics. The fungi were determined on the basis of morphological and metabolic characteristics according to Egorova (1986) and Rafai (1969). Algae were kindly determined by Professor E.A. Shtina in water cultures with nutrient medium and on over-growth glasses on the soil plates.

3   Results and discussion

Microeukaryotes (fungi, yeasts, algae) are more tolerant of heavy metals than prokaryotes (bacteria, cyanobacteria, streptomycetes). Lists of the frequently occurring species of fungi and algae in the upper organic horizons of soils are given in Tables 5.1 and 5.2. In unpolluted soil, there were 28 fungal species with an occurrence of 30–60%. In the buffer zone, the species diversity of fungi remained the same, whereas in the impact zone the species number decreased to 19. In the central zone, only 15 species were present, with the fungal flora being dominated by *Paecilomyces farinosus*, *Philophora melinii*, *Penicillium simplicissimum*, *Trichoderma koningii*, *Trichoderma aureoviride*, *Mucor griseo-cyanus* and sterile mycelia. These fungi appear to be tolerant of elevated copper

Table 5.1. Changes in the taxonomic diversity of fungi in soils under the impacts of emissions from the 'Severonikel' enterprise. Content of heavy metals in soils (mg kg^{-1}): background zone Cu, 40, Ni, 80; buffer zone Cu, 200–400, Ni, 500–1000; impact zone Cu, 600–1000, Ni, 1700–2600; epicentre Cu, 3000–6000, Ni, 7000–15,000. Dotted lines indicate that the maximum concentration of heavy metals preventing growth has not been established.

Species	Zone of observations			
	Background	Buffer	Impact	Epicentre
Acremonium spp.	←—————→			
Aspergillus niger	←——————————————————→			
Aspergillus spp.	←——————————————————→			
Aureobasidium pullulans	←········			
Chaetomium spp.	←—————→			
Mortierella ramanniana	←—————→			
Mucor griseo-cyanus	←——————————————————→			
Mucor spp.		←——————————→		
Paecilomyces farinosus	←——————————········			
Philophora melinii		←——————————→		
Penicillium canescens	←———————————→			
Penicillium corylophilum	←——————————————————→			
Penicillium implicatum		←——————————→		
Penicillium frequentans	←—————→			
Penicillium notatum	←—————→			
Penicillium simplicissimum	←——————————————········			
Penicillium spinulosum	←—————→			
Penicillium sp. 1	←——————————————········			
Penicillium sp. 2	←———————————········			
Penicillium steskii	←—————→			
Rhizopus nigricans	←———————————→			
Rhodotorula glutinis	←——————————————········			
Sterile mycelium	←——————————————········			
Trichoderma aureoviride	←——————————————————→			
Trichoderma lignorum	←—————→			
Trichoderma viride	←—————→			
Trichoderma koningii	←——————————————————→			
Verticillium spp.	←———————————→			

concentrations in the soil. In contrast, there were no representatives of the genera *Chaetomium*, *Acremonium*, *Verticillium*, *Mortierella* and *Rhizopus*, all of which were present in the unpolluted soil. *Mucor* sp. and *Philophora melinii*,

Table 5.2. Changes in the taxonomic diversity of algae in soil under impact of emissions of 'Severonikel' enterprise. See Table 5.1 for details.

	Zone of observations			
Species	Background	Buffer	Impact	Epicentre
Cyanophyta				
Nostoc linckia	←——→			
N. muscorum	←——→			
Phormidium spp.	←——→			
Synechococcus elongatus	←——→			
Xanthophyta				
Botrydiopsis arhiza	←————	————→		
Characiopsis minuta	←——→			
Characiopsis saccata	←——→			
Heterothrix stichococcoides		←——→		
Pleurochloris spp.	←——→			
Polyedriella helvetica	←——→			
Chlorophyta				
Bracteacoccus minor	←————	————	————	————→
Chlamydomonas elliptica	←————	————	————	————→
Chlorella mirabilis	←————	————	————	————→
Chlorella vulgaris	←————	————	————	————→
Chlorococcum spp.	←————	————	————	————→
Chlorhormidium flaccidum	←————	————→		
C. flaccidum var. *nitens*	←————	————→		
Chlorhormidium dissectum	←——— ···			
Closterium pusillum	←——— ···			
Coccomyxa solorinae	←————	————	————	————→
Cosmarium decedens	←——→			
Cylindrocystis brebissonii	←————	————	···	
Desmococcus vulgaris	←————	————	————	————→
Leptosira spp.	←————	———— ···		
Microthamnion strictissimum		←——→		
Myrmecia bisecta	←——→			
Myrmecia incisa		←——→		
Stichococcus bacillaris	←——→			
Stichococcus minor	←————	————	————	————→
Bacillariophyta				
Eunotia spp.			←———	——→
Pinnularia borealis	←————	————	——→	

identified in the polluted soil (centre and impact zones), are not typical of the less polluted soils. When copper concentrations in the soil reach 1600–3000 mg kg^{-1} and nickel concentrations reach 3500–6000 mg kg^{-1}, the dominant fungi belong to the genus *Penicillium*. These are the most tolerant fungi to copper and nickel.

Newby and Gadd (1985) also noted differences in the sensitivity to copper of different species of fungi. Of the fungi that they investigated, the greatest sensitivity to copper was shown by *Chaetomium globosum*, and the least by *Phoma violace* and *Penicillium ocho-chloron*. They found that the toxicity of copper to filamentous fungi increased at lower pH values, as was also observed in the impact zone of the Severonikel smelter affected by sulphur dioxide. A reduction in the length of the fungal mycelium, reduced fungal biomass and up to 75% reductions in soil respiration rates have been attributed to copper concentrations in the soil of 1000 mg kg^{-1} (Nordgren *et al.*, 1985). The metal pollution of the soil has more impact on the growth and development of fungi than the humidity or organic content of the soil. Some fungi have well-developed tolerance mechanisms, particularly through the immobilization of toxic elements by the formation of chelates between metals and fungal meta-bolites or by the bioaccumulation of metallic ions by the mycelium (Gadd and Griffiths, 1978; Gadd, 1990; Evdokimova and Mozgova, 1992). These mechanisms enable fungi to tolerate the acidic soil conditions in the centre of emissions from non-ferrous metallurgy enterprises.

The number of algal species also decreased towards the smelter, with 27 species in unpolluted soil and only nine close to the smelter. The most tolerant were the unicellate green algae, especially species of the genera *Chlamydom-onas*, *Chlorococcum*, *Bracteacoccus* and *Chlorhormidium*. These algae generate a polysaccharide-rich slime that protects them against toxic metals (Sorentino, 1985). The reduction in the number of algal species occurred through the disappearance of blue-green and yellow-green algae. However, some species were only present in the heavily polluted soils, namely *Heterothrix sticho-coccoides*, *Microthamnion strictissimum*, *Myrmecia incisa* and *Eunotia* spp.

The bacteria which are most sensitive to elevated soil concentrations of copper and nickel are the neutrophiles – the nitrogen-fixing and nitrification bacteria (Table 5.3). Saprophytic non-spore-forming bacteria are also sensitive. The latter belong to the r-strategy group of species, having weak competitive abilities and high rates of growth (Andrews and Harris, 1986; Zvyagintsev, 1994). Species belonging to the ecological group characterized by the K-strategy, including many bacilli and streptomycetes, are more tolerant as they are able to generate stable forms (spores, cysts).

The homeostasis of the microbial community in podzolic soil was disrupted at soil copper concentrations in excess of 300–400 mg kg^{-1} and nickel concentrations in excess of 600–700 mg kg^{-1} (Evdokimova, 1995a, b). This means that the most sensitive groups were affected when pollutant levels reached ten times the background levels. The most tolerant groups were

Table 5.3. Changes in the taxonomic diversity of bacteria in soil under impact of emissions of the 'Severonikel' enterprise. See Table 5.1 for details.

Species (genus)	Zone of observations			
	Background	Buffer	Impact	Epicentre
g. *Clostridium*	←————————→			
g. *Nitrosomonas*	←————————→			
g. *Nitrobacter*	←————→			
Pseudomonas denitrificans	←————————————————→			
Pseudomonas fluorescens	←————————————→			
Pseudomonas spp.	←————————————→			
Corynebacterium insidiosum	←————————————→			
Streptomyces albus	←————————————————→			
Streptomyces griseus	←————————————→			
g. *Arthrobacter*	←————————————————→			
g. *Rhodococcus*	←——————————————————→			
g. *Nocardia*	←————————————→			
g. *Cytophaga*	←————————————→			
Bacillus licheniformis	←————————————————→			
Bacillus megaterium	←————————————————→			
Bacillus subtilis	←————————————→			
Bacillus spp.	←——————————————————————→			

affected when levels exceeded 100 times the background concentrations. The soils 7–10 km downwind of the smelter, which have been impacted for several decades (40–50 years), have lost whole physiological groups of bacteria, including nitrogen-fixing, nitrifying, cellulolytic bacteria and cyanobacteria.

4   Conclusion

The pollution of soils by heavy metals and sulphur compounds has caused a decrease in the species diversity of soil microorganisms. The diversity of the dominant species of non-spore-forming saprophytic bacteria, algae and fungi were all reduced. The homeostasis of the microbial community in the podzolic soil has been changed by copper concentrations of 300–400 mg kg^{-1} and nickel concentrations of 600–700 mg kg^{-1}. Generally, the microeukaryotes, particularly the fungi, are less sensitive to heavy metals (Cu, Ni, Co) than prokaryotes. Whole functional groups of bacteria and cyanobacteria have been lost from soils exposed to high pollutant loads for the last 40–50 years. The disappearance of some sensitive physiological groups of microorganisms and the decrease in the diversity of less sensitive trophic and taxonomic groups has caused changes in the ecosystem functions of the soil microorganisms and has resulted in changes in the nutrient status of the soils. This provides clear evidence of changes in the ecosystems that are not readily apparent by standard assessment techniques and indicates that changes in the microflora of the soil could be an important additional tool for monitoring the impacts of pollutants on forest ecosystems.

Acknowledgements

This work was supported by the Russian Fund of Basic Investigations (Grant [1] 04 – 48508).

References

Andrews, J.H. and Harris, R.F. (1986) r- and K-selection and microbial ecology. *Advances in Microbial Ecology* 9, 99–147.

Atlas, R.M. (1984) Diversity of microbial communities. *Advances in Microbial Ecology* 7, 1–48.

Baath, E., Frostegard, A. and Fritze, H. (1992) Soil bacterial biomass, activity, phospholipid fatty acid pattern, and pH tolerance in an area polluted with alkaline dust deposition. *Applied and Environmental Microbiology* 58, 4026–4031.

Egorova, L.N. (1986) *Soil Fungi of the Far East*. Nauka, Leningrad (in Russian).

Evdokimova, G.A. (1995a) Effects of pollution on microorganisms biodiversity in Arctic ecosystems. In: Anon. (ed.) *Global Change and Arctic Terrestrial Ecosystems*. European Community, Brussels and Luxembourg, pp. 315–320.

Evdokimova, G.A. (1995b) *Ecological and Microbiological Foundations of Soil Protection in the Far North*. Kola Science Centre, Apatity (in Russian).

Evdokimova, G.A. and Mozgova, N.P. (1992) Accumulation of copper and nickel by soil fungi. *Microbiology* 60, 550–554.

Gadd, G.M. (1990) Heavy metals accumulation by bacteria and other microorganisms. *Experientia* 8, 834–840.

Gadd, G.M. and Griffiths, A.T. (1978) Microorganisms and heavy metals toxicity. *Microbial Ecology* 4, 303–317.

Holt, J. and Zavarzin, G.A. (eds) (1980) *The Shorter Bergey's Manual of Determinative Bacteriology*. Mir, Moscow.

Lockwood, J. and Pimm, S.L. (1994) Species: would any of them be missed? *Current Biology* 4, 455–457.

Newby, P.J. and Gadd, G.M. (1985) Assesment of heavy metal toxicity towards fungi. In: Allan, R.J. and Nriagu, J.O. (eds) *Heavy Metals in the Environment*, Vol. 2. CEP Consultants, Edinburgh, pp. 162–164.

Nordgren, A., Bååth, E. and Söderström, B. (1985) Soil microfungi in an area polluted by heavy metals. *Canadian Journal of Botany* 63, 448–455.

O' Donnell, L.G., Goodfellow, M. and Hawksworth, D.L. (1994) Theoretical and practical aspects of the quantification of biodiversity among microorganisms. *Philosophical Transactions of the Royal Society of London* 345, 65–73.

Pennanen, T., Frostegarg, A., Fritze, H. and Bååth, E. (1996) Phospholipid fatty acid composition and heavy metal tolerance of soil microbial communities along two heavy metal-polluted gradients in coniferous forest. *Applied and Environmental Microbiology* 62, 420–428.

Rafai, M.A. (1969) A revision of the genus *Trichoderma*. *Mycological Papers* 116, 1–56.

Sorentino, C. (1985) Copper resistance in *Hormidium fluitens* (Gay) Heering. *Phycologia* 24, 366–368.

Torsvik, V., Salte, K., Sorheim, R. and Goksoyr, J. (1990) Comparison of phenotypic diversity and DNA heterogeneity in a population of soil bacteria. *Applied and Environmental Microbiology* 56, 776–781.

Tresner, H.D., Bacus, M.P. and Curtis, I.T. (1954) Soil microfungi in relation to the hardwood forest continuum in Southern Wisconsin. *Mycologia* 46, 314–317.

Zvyagintsev, D.G. (1994) Vertical distribution of microbial communities in soil. In: Ritz, K., Dighton, J. and Giller, K.E. (eds) *Beyond the Biomass: Compositional and Functional Analysis of Soil Microbial Communities*. John Wiley & Sons, New York, pp. 29–37.

Air Pollution Impacts on Subarctic Forests at Noril'sk, Siberia

6

V.I. Kharuk

Sukachev Institute of Forest, Krasnoyarsk, Russian Federation

The Noril'sk mining industrial complex (69°N, 88°E) is having a dramatic impact on the forest–tundra transition area in the northern part of Siberia. The main pollutant, sulphur dioxide, has caused forest damage and mortality at distances up to 200 km downwind from the smelters. The sensitivity of trees and shrubs to pollutants decreases in the following order: larch, spruce, birch and willow. The soil sulphur content varies within the range 30–540 kg ha^{-1} in the upper 0–2 cm, and from 100 to 250 kg ha^{-1} in the 0–20 cm layer. Heavy metals (copper, nickel and cobalt) in the upper soil horizons (0–10 cm) exceed background levels by a factor of 10–1000 at distances of up to 30 km from the smelters. A dramatic increase in forest mortality in the late 1970s and the early 1980s was attributed to the construction of taller stacks which facilitate pollution transfer over longer distances. The Noril'sk phenomenon is one of the greatest ecological catastrophes caused by man in the subarctic zone.

1 Introduction

Tundra and subarctic forests occupy 20% of the Siberian territory. The zone of sparse forests is about 50 million hectares (Mha) with a volume of wood estimated at *c.* 1.3 billion m^3. The northern tree-line in Siberia is formed by Gmelinii larch (*Larix gmelinii* Rupr.) and Siberian larch (*Larix russica* (Endl.) Sabine). The subarctic forests have a valuable ecological function in mitigating biosphere processes, and the growth dynamics and position of the northern tree-line are considered to be indicators of climatic trends, as temperature is a limiting factor in this region. The extreme climatic conditions make plant communities of the far north highly sensitive to anthropogenic impacts. In Siberia, subarctic forests have been heavily affected by the Noril'sk mining industrial complex for more than half a century.

Noril'sk (69°N, 88°E), the world's biggest city north of the polar circle (c. 180,000 residents), was founded in 1935. The Noril'sk area contains one of the biggest deposits of nickel (one third of the world reserves), copper (c.12% of world reserves), cobalt, gold and platinoids (c. 15% of world reserves, and 99.98% of the reserves of the former USSR). The Noril'sk complex produces 30% of the world's nickel and more than half of the world's platinoids. As the metals in the ore are bound with sulphur, the contents of sulphur and metals in the ore are correlated. The sulphur content in the ore varies from 2.3% ('poor ore') to 28% ('rich ore'). Under current economic pressures, ore with a high metal–sulphur content is being used for smelting. The annual output of the primary pollutant, SO_2, is about 1.7 Mt, or more than 20% of the total SO_2 emissions in Russia (Anon., 1995). Not all the ore that is mined is refined at Noril'sk. For example, ore is exported to the smelters of the Kola Peninsula (Rigina and Kozlov, this volume).

The history of forest destruction around Noril'sk goes back to the beginning of the construction of the industrial complexes. In the 1930s and at the beginning of the 1950s, the forests were used as a source of wood. A buffer, 3–4 km in width, was cut around prison camps (prisoners were used extensively for the construction of the Noril'sk industry). Pollutant emissions (SO_2, N_xO_x and heavy metals) started at the beginning of the 1940s; in 1942, in the middle of World War II, the Noril'sk industrial complexes provided the first nickel and copper for military needs. The stacks of the main smelter have the following heights: 138 m, 150 m (constructed in the 1940s), 180 m (the 1950s) and 250 m (constructed in 1980). The earliest available data on forest decline caused by pollution in the Noril'sk area is from the late 1960s. The area of dead forest was estimated to be approximately 5000 ha. However, local residents consider that the degradation began earlier, in the mid-1950s. Forests in the area are continuing to decline today.

2   Materials and methods

The forests of the Noril'sk area are concentrated mainly in the Noril'sk valley and around lakes and rivers. The Noril'sk valley is a tectonic depression 50–100 m above sea level with many bogs and small lakes. It is 150–170 km long and 20–60 km wide (Fig. 6.1). As almost all forest stands along the valley are now dead, local people refer to it as a 'valley of death'. In the west, the valley borders the Lontokoy ridge, with maximum heights of 600–700 m. Its easterly border is the Putorana Plateau, with heights of 800–1000 m. The forest stands are formed by larch (the dominant species), Siberian spruce (*Picea obovata* Ledeb.), silver birch (*Betula pendula* Roth.) and a variety of willow species. The stands are mainly old (200–300 years) and of low density (0.4 average, up to 0.8 in river valleys). The soils are primarily very poorly drained brown, cryaquept and tundra histosols. The average depth of soil thaw during

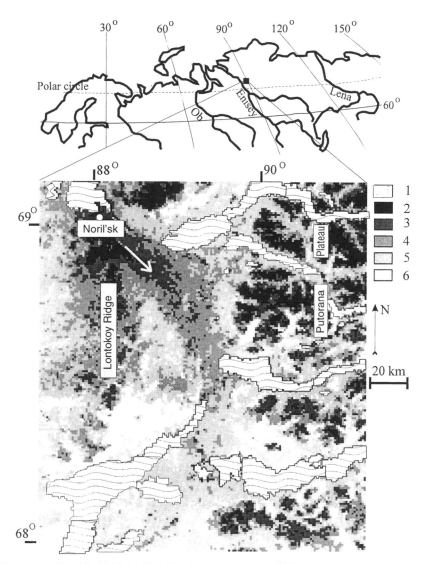

Fig. 6.1.   Classified NOAA/AVHRR image of the Noril'sk region. The arrow indicates the direction of the prevailing winds. Image taken in summer 1997. 1, water; 2, rocks; 3, industrial barren; 4, heavily damaged stands; 5, moderately damaged stands; 6, slightly damaged stands.

summer is 0.4–0.6 m. The forest floor is composed of thick pillows of moss and lichen. Precipitation ranges between 400 and 500 mm annually. Average daily air temperatures above +10°C occur from the middle of July until late August and the growing season is approximately 60 days.

Permanent test plots were installed along the pollution gradient, up to a distance of 200 km from the source. Tree health was estimated by scaling the visual symptoms of defoliation, discoloration and changes in crown architecture, with the European scale being used for this purpose (Anon., 1985). The concentrations of various pollutants in the foliage, forest floor and soil were determined. Soil toxicity was assessed using the following indices: $(Ca^{2+} + Mg^{2+})/Cu^{2+}$ and $(Ca^{2+} + Mg^{2+})/Ni^{2+}$. At values below 10 and 5, respectively, soils are considered to be toxic (Evdokimova 1995). Satellite (Landsat MSS, NOAA/AVHRR, Russian satellite) images and archival forest inventory data were used to evaluate forest decline dynamics. The zone of vegetation decline was determined using the NDVI (normalized difference vegetation index), as this index correlates with biomass (Kharuk *et al.*, 1992; Goward *et al.*, 1993).

3   Results and discussion

3.1   Air and soil pollution

The dynamics of SO_2 emissions at Noril'sk are presented in Fig. 6.2. Peak SO_2 emissions, 2.5 Mt year^{-1}, occurred in the 1980s (according to another source, the maximum output in that period was 2.8 Mt year^{-1}). In the 1990s, annual SO_2 emissions were approximately 1.7 Mt. The occurrence of abnormally high SO_2 emissions resulting from technology violations is unknown but is likely to be an important factor in the extent and development of forest damage. The output of N_xO_x is about 20,000 t year^{-1}. The reported data for heavy metal (Ni, Cu, V) emissions are *c.* 4000, 1800 and 65 t year^{-1}, respectively (Nilsson *et al.*, 1998).

The smoke plume from the Noril'sk industrial complex is carried in a south-east direction along the Noril'sk basin (Fig. 6.1). The concentrations of primary pollutants in the soil along this pollution gradient are shown in Fig. 6.3. The soil sulphur content varies from 30 to 540 kg ha^{-1} for the upper 0–2 cm, and from 100 to 250 kg ha^{-1} for the 0–20 cm layer. Despite the heavy pollution levels, there have not been dramatic changes in soil pH. Even in the central impact zone, pH values never fall below 4.4; normal values lie between 5.3 and 7.0. Snow-water pH is neutral or slightly alkaline. The process of acidification is mitigated by the high buffering capacity of the soil and by CaO emissions from the local cement industry.

Heavy metal (Cu, Ni, Co) contents in the upper soil horizon (0–10 cm) exceed the background level by a factor of 10–1000 within approximately

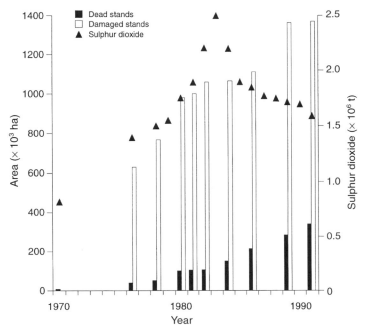

Fig. 6.2. Sulphur dioxide annual output and the dynamics of stand damage and mortality.

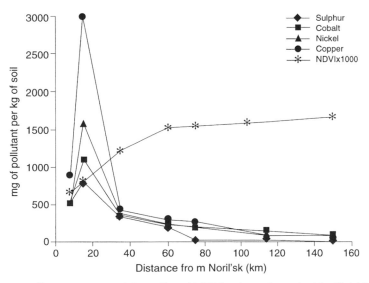

Fig. 6.3. Pollutant content of the soil and NDVI values along the Noril'sk Valley.

30 km of the smelters (Fig. 6.3). Maxima on the graphs correspond to the areas where the plume 'touches' the ground. Judging by the indices described above, soils are toxic within a radius of 25–30 km of the smelters; for example at a distance of 4 km, the values of the indices are in the range of 0.4–1.5 (they should normally be 100–200). The NDVI (the vegetative index derived from an NOAA/AVHRR image taken in 1997) has minimal values within approximately the same distance (Fig. 6.3), indicating that even grass communities are heavily depressed in the area. The subsequent increase in the NDVI values can be explained by: (i) the development of grass communities in areas where larches have died; (ii) birch and willow species; and (iii) surviving larch stands (mainly in local topographic depressions).

3.2   Effects on forest stands

Forest damage has been detected at a distance of up to 200 km from the smelters, and forest mortality at a distance of 80–100 km (Kharuk *et al.*, 1996). The data derived from NOAA/AVHRR images (Fig. 6.1) illustrate the scale of the forest damage. A mountain ridge (average heights 800–900 m) limits the spread of pollution to the south. However, lichens beyond the ridge show signs of damage at a distance of 200–250 km (Kharuk, 1992). The analysis of a time series of space images has revealed a dramatic increase in forest decline at the end of the 1970s and the beginning of the 1980s (Kharuk *et al.*, 1994). Growth increment measurements also show an increase in the rate of forest mortality to 5–15 times normal rates (Ivshin, 1993). This increase in damage rate is correlated with the construction of taller (250 m) stacks. The 'effective height' of these new stacks is even greater as they were constructed on a hill (*c.* 200 m higher than the surrounding surfaces). Figure 6.2 indicates the official emission rates for sulphur dioxide emission and forest mortality. In the mid-1990s, forests had been killed in an area extending to 400,000 ha, and damaged over *c.* 1.5 Mha. According to Russian forest inventory rules, the forests are only classified as such when the stand density exceeds 0.3, even though stands in the subarctic are expected to have a density of 0.1 (e.g. Shiyatov, 1985). Consequently, the official statistics for forest mortality underestimate considerably the extent of the damage.

3.3   The symptoms of damage

Typical symptoms of tree damage include discoloration, defoliation, changes in crown shape, reductions in increment and (for spruce) reductions in needle retention (to 1–3 years, compared with 10–12 years under normal conditions). In spruce, low branches are less damaged, possibly as a result of protection by snow. Leaves of broad-leaved deciduous trees and shrubs (birch, willow) are covered with yellow-brown spots on the margins and between veins. In the zone of chronic SO_2 impacts, discoloration due to the loss of chlorophyll is typical. Acute injury by the pollution plume causes red-brown necrosis of larch needles. After such a 'shock', larch can survive due to regrowth of the needles from brahiblasts or 'dormant buds'. These compensatory needles are bigger (up to 3–4 times) and abnormal in shape (they are often curled). Another compensatory reaction is the intensive growth of needles along the stem and primary branches. The concentrations of some biogenic elements (Mg, K, Ca, Fe) in larch needles is 1.3–5.0 times higher in areas affected by pollution than in adjacent unaffected areas.

Stand vigour correlates better with needle sulphate ion concentrations than with sulphur deposition values. In the impact zone, the concentration of sulphate ion in the needles is 3–5.5 mg g^{-1}, as compared with a background level of 0.5–0.8 mg g^{-1} (Kharuk et al., 1996). The impact of heavy metals is limited to an area within about 30 km distance of the smelters, although forest damage is measurable at a distance of 200 km. Hence, the primary cause of forest decline is considered to be SO_2 or its derivatives in gaseous or aerosol form.

There are no data on SO_2 concentrations in ambient air along the pollution gradient, except in the city of Noril'sk. Reported sulphur dioxide concentrations in the city are between 0.1 and 0.2 mg m^{-3}, with peak concentrations above 20 mg m^{-3}. This exceeds, by a factor of 40, the critical concentrations for human health (0.05 mg m^{-3}). Nikolaevsky (1979) and Treshow (1984) state that the critical levels of SO_2 for larch species are 0.15 mg m^{-3} for peak values, 0.03 mg m^{-3} for the maximum average daily concentrations, and 0.01 mg m^{-3} for daily concentrations over the entire growing season.

4   Factors affecting the sensitivity/resistance of trees to pollutants

In areas experiencing acute effects from the plume, even the most resistant species (Salix spp.) are damaged. Outside this zone, the level of damage depends on the prevailing winds, topography and soil fertility. Forest stands on sheltered slopes and stands in the adjacent river valleys are considerably less affected. Usually, the healthiest stands are found on fertile soils, which have organic matter contents 1.5–4 times higher and nitrogen contents 2.8–8 times higher than soils supporting more heavily damaged stands. Among a number of other factors which affect tree/forest sensitivity/resistance, the most noticeable are severe climatic conditions, solar radiation intensity, stand density and air humidity/fog/rains. Solar irradiation during the vegetation period in the subarctic zone is higher than in equatorial regions (Budyko, 1962), although the significance of this is still uncertain. A factor that may be important is that the continuous daylight means that pollutant uptake can occur without respite, so there is no opportunity for the plants to mobilize repair mechanisms.

The sensitivity of trees and shrubs to pollutants decreases from larch (most sensitive), through spruce and birch to willow (least sensitive). This contrasts with suggestions that larch are more tolerant than spruce (Treshow, 1984). We found that larch was more sensitive in the areas with intensive chronic pollution impact (e.g. the 'valley of death'). In this case, spruce may have some advantages due to the greater resistance of its needle cuticles to SO_2 or its derivatives. On the other hand, in areas experiencing very high SO_2 peaks, larch shows more tolerance. For example, small stands in local topographic depressions growing about 5 km upwind from the smelters are formed only by larch. Needles of these larches are periodically affected by SO_2, involving needle damage and mortality. However, during the periods between peak concentrations, needles regenerate, and trees show satisfactory increment and germination. Evidently, this kind of survival depends on the frequency of the extreme concentrations and warrants further investigation.

5   Natural causes of forest decline

In addition to pollution, there are several natural factors causing the decline of subarctic forests. These include stand ageing, poor regeneration, insects and fungal diseases. More than 80% of the forests in the Noril'sk area are mature or old-growth, i.e. there is a natural process of forest decline. Larch regeneration is limited by the thick layer of lichens on the ground and by moss 'pillows'. These 'pillows' also act as insulation, aggravating the thermal conditions in the root zone by decreasing the depth of thawing, and excluding roots from deeper soil horizons. Wild fires naturally promote the regeneration process, as they normally burn down to mineral soil. On burned areas, regeneration is 10–20 times greater than on unburned areas, and seedling counts can reach $10,000–20,000$ ha^{-1}. Because of the harsh climatic conditions, insect populations are generally low, and their impact is considered to be negligible. Stem fungal diseases affect 6–8% of trees. Some fungi also affect needles, causing their partial yellowing. In both cases there is no evident correlation with forest vigour. Dendrochronological data have been used to identify climate as a partial cause of forest decline (E.A. Vaganov, personal communication). Non-pollution factors have been estimated to account for 10–20% of the forest damage.

6   Conclusion

Air pollution-induced forest decline in Siberia is the greatest anthropogenic ecological catastrophe in the subarctic zone. Stands up to 80–100 km downwind of the smelters at Noril'sk are dead, and they are damaged up to 200 km from the smelters. Data obtained by local foresters indicate that forest mortality is progressing downwind at a rate of about 2 km year^{-1}. The primary cause of forest decline in the Noril'sk area is sulphur dioxide or its derivatives in gaseous or aerosol forms; current SO_2 emissions are about 1.7 Mt annually. In comparison, the total emissions of SO_2 in the more intensively studied Kola Peninsula are only 0.6 Mt year^{-1} (Lukina and Nikonov, 1996). The well-documented studies of similar sites in North America (e.g. Copper Hills, USA and Sudbury, Canada) suggest considerably less impact on the environment (Hutchinson and Whitby, 1977). Clearly the problems in the subarctic forests are very different from forest declines elsewhere (Schütt, 1988), providing the opportunity for further studies. As pollution impacts have occurred for more than half a century, possible investigations of natural selection and adaptation in larch are also of interest. Another aspect of the problem lies in the biophysical links between climate change and regional anthropogenic impacts on the radiation balance. The latter is due to: (i) changes in the winter/summer albedo of the territory; and (ii) attenuation of solar radiation by industrial aerosols. This is of special interest as we have identified a climate-induced forest expansion into the tundra zone in northern Siberia (outside the direct impact zone of the Noril'sk smelters).

References

Anonymous (1985) *Manual for the Assessment and Monitoring of Air Pollution Effects in Forests*. International Cooperative Programme for the Assessment and Monitoring of Air Pollution Effects in Forests, Hamburg.

Anonymous (1995) *Environment of Russia, 1995*. Ecos, Moscow, pp. 82–83 (in Russian).

Budyko, M.I. (1962) Polar ice and climate. *Izvestiya of USSR Academy of Sciences* (0) 6, 23–31 (in Russian).

Evdokimova, G.A. (1995) *Ecological and Microbiological Foundations of Soil Protection in the Far North*. Kola Science Centre, Apatity (in Russian).

Goward, S.N., Dye, D.G., Turners, S. and Yang, J. (1993) Objective assessment of the NOAA global vegetation index data product. *International Journal of Remote Sensing* 14, 3365–3394.

Hutchinson, T.C. and Whitby, L.M. (1977) The effect of acid rainfall and heavy metal particles on the boreal ecosystem near the Sudbury smelting region of Canada. *Water, Air, and Soil Pollution* 7, 421–438.

Ivshin, A.P. (1993) Norilsk air pollution impact on spruce-larch stands vigour. PhD Abstract, Ekaterinburg (in Russian).

Kharuk, V.I. (ed.) (1992) Monitoring of Forest Stands and Soils of the Norilsk Region. Annual report, Krasnoyarsk, unpublished manuscript (in Russian).

Kharuk, V.I., Alshansky, A.M. and Yegorov, V.V. (1992) Spectral characteristics of vegetation cover: factors of variability. *International Journal of Remote Sensing* 17, 326–327.

Kharuk, V.I., Winterberger, K.C., Tzibulsky, G.M. and Yahimovich, A.P. (1994) Pollu-
    tion induced pretundra forest decline. In: *Proceedings of the 1994 Society of
    American Foresters/Canadian Institute of Forestry Convention, Anchorage, Alaska.*
    Society of American Foresters, Bethesda, Maryland, pp. 442–447.
Kharuk, V.I., Winterberger, K.C., Tzibulsky, G.M., Yahimovich, A.P. and Moroz, S.N.
    (1996) Pollution induced damage of Norilsk valley forests. *Ecologiya [Ecology,*
    Ekaterinburg] (0) 6, 424–430 (in Russian).
Lukina, N.V. and Nikonov, V.V. (1996) *Biogeochemical Cycles in the Northern Forests
    Subjected to Air Pollution. Part 1.* Kola Science Centre, Apatity (in Russian).
Nikolaevsky, V.S. (1979) *The Biological Basis for Gas-Resistance of Plant Species.* Nauka,
    Novosibirsk (in Russian).
Nilsson, S., Blauberg, K., Samarskaia, E. and Kharuk, V. (1998) Pollution stress of
    Siberian forests. In: Linkov, I. and Wilson, R. (eds) *Air Pollution in the Ural Moun-
    tains.* Kluwer, Dordrecht, pp. 31–54.
Schütt, P. (1988) Forest decline in Germany. In: Hertel, G. (technical coordinator)
    *Proceedings of the US/FRG Research Symposium: Effects of Atmospheric Pollutants in
    the Spruce-fir Forests in the Eastern United States and the Federal Republic of Germany,
    19–23 October, Burlington, Vermont.* General Technical Report NE-120, US Depart-
    ment of Agriculture Forest Service, Broomall, Pennsylvania, pp. 87–88.
Shiyatov, S.G. (1985) Definition of upper forest line. In: Gorchakovsky, P.L. (ed.) *The
    Ural Vegetation and Its Anthropogenic Changes.* Academy of Sciences, Sverdlovsk,
    pp. 32–58 (in Russian).
Treshow, M. (ed.) (1984) *Air Pollution and Plant Life.* John Wiley & Sons, Chichester.

Landscape Degradation by Smelter Emissions near Sudbury, Canada, and Subsequent Amelioration and Restoration

7

K. Winterhalder

Department of Biology, Laurentian University, Sudbury, Canada

A century of sulphur dioxide fumigation, copper and nickel particulate deposition, fire, soil erosion and enhanced frost action has created 17,000 ha of barren land and 72,000 ha of stunted, open birch–maple woodland in the Sudbury area. The primary factor limiting plant colonization is the acidic, aluminium-, copper- and nickel-toxic properties of the soils, although certain plant species have developed genetically based metal tolerance. In the revegetation programme, manual surface application of ground limestone, with or without an accompanying fertilizer and/or grass–legume seed application, leads to immediate colonization by woody species including birch, aspen and willows, and more than 3000 ha have been treated in this way by the Regional Municipality of Sudbury since 1978. Native coniferous species have also been planted in groups to form a seed source for future colonization.

1   Introduction

The mining and smelting region of Sudbury (46°30′N, 80°00′W) lies at a mean altitude of 300 m on the southern edge of a geological feature known as the Precambrian, Canadian or Laurentian Shield. The landscape is rugged, with numerous glacially scoured rock outcrops and narrow valleys. The Wisconsin glaciers retreated about 10,000 years ago, and soils were formed on glacial till of varying depth, and on glacio-fluvial material. The elliptical Sudbury Basin, around the edge of which nickel and copper ores are mined, is

believed by many to have been formed by meteorite impact nearly 2 billion
years ago (Pearson and Pitblado, 1995).

Sudbury lies near the northern edge of the vegetation zone referred to by
Nichols (1935) as the Hemlock–White Pine–Northern Hardwoods Region,
and by Rowe (1959) as the Great Lakes–St Lawrence Forest Region. Sudbury
was once characterized by extensive stands of red and white pine *(Pinus
resinosa* Ait. and *Pinus strobus* L.), while to the south, white pine formed an
admixture with tolerant hardwoods such as sugar maple *(Acer saccharum*
Marsh.) and yellow birch *(Betula alleghaniensis* Britt.). The large white pine
stumps now found on bare, stony slopes (Fig. 7.1) of the Sudbury pollution
zone, and the vestiges of white cedar *(Thuja occidentalis* L.) stumps that cover
many hectares of barren peat (Fig. 7.2), suggest a mosaic of pine forests on the
slopes and cedar swamps in many of the depressions.

Lumber operations began in 1872, and the larger red and white
pines were cut and floated down rivers to Georgian Bay and Lake Huron,
then rafted to sawmills in the northern United States. By the turn of the
century, black and white spruce *(Picea mariana* (Mill.) BSP and *Picea glauca*
(Moench) Voss), balsam fir *(Abies balsamea* (L.) Mill.), and later jack pine
(Pinus banksiana Lamb.) were also being harvested, and more than 11,000
men were employed in the mills and in the forest around Sudbury. In the wider
Sudbury area, lumber operations continued to be the dominant industry
as late as 1927, despite the emergence of mining after 1886 (Winterhalder,
1996).

Fig. 7.1.   Barren, stony slope with white pine stumps.

Fig. 7.2. Barren peatland with white cedar stumps.

2 Degradation processes in the Sudbury landscape

Early selective logging for red and white pine made a minimal contribution to environmental degradation, and rapidly growing successional species such as white birch (*Betula papyrifera* Marsh.) and trembling aspen (*Populus tremuloides* Michx.) quickly colonized the gaps. Even the less-selective logging that came later, for railroad ties or sleepers, locomotive fuel, pit timbers and pulpwood, would not normally have led to the permanent denudation of the landscape. The construction of the Canadian Pacific Railway resulted in increased timber cutting and a greater frequency of fire. Sparks from the wood-burning loco-motives contributed to the fires and prospectors were said to have exposed the bedrock under the duff cover by burning. As logging became less selective, a greater proportion of the timber was removed, and the slash left behind created an ideal fuel for forest fires. Later, sulphur dioxide damage led to foliage death, creating tinder.

2.1 Open-bed roasting

When mining in the Sudbury area began in 1885, the ore was sent elsewhere for processing. In 1888, the first roast yard and smelter were set up in Copper Cliff. At this time, roasting, which involves heating the ore in the presence of air so that the more reactive sulphides are oxidized, was carried out by piling crushed ore on beds of wood in the open (Fig. 7.3), covering them with the finer material to prevent open flames, and igniting. At first, heat was provided by the burning wood, then the iron sulphide (pyrrhotite) in the ore began to burn. After burning for 2 or more months (Fig. 7.4), the 'calcined' ore was loaded on to rail cars and transported to furnaces for smelting and 'conversion'. Near the

Fig. 7.3. O'Donnell open roast yard, with crushed ore piled on beds of wood (Inco archives).

Fig. 7.4. O'Donnell roast yard in operation (Inco archives).

roast yards, even the smallest timber was removed as fuel. In 1916, the 11 roast yards were replaced by a very large and highly mechanized yard at O'Donnell, some distance west of Sudbury. Open-bed roasting was abandoned in 1929.

Open-bed roasting is often blamed for the widespread degradation of the Sudbury landscape. However, although the dense sulphur dioxide fumes emitted from roast beds at ground level killed plants and acidified soils in their path, the effects of the roast beds on vegetation were neither as severe nor as permanent as popularly believed. Rather, the permanent and widespread poisoning of the soil was caused later by fumes from the blast furnaces, which contained copper and nickel particles as well as sulphur dioxide. It seems that the roast beds produced large amounts of sulphur dioxide, but relatively little particulate metal.

In 1946, 17 years after the closure of the O'Donnell roast yard, air photographs show it surrounded by a patchy cover of shrubs. By 1959 pioneer trees such as poplars and birches had begun to colonize, while by 1973 there was mosaic of forest cover, but still with intermittent openings. Currently, the birch forest grows right to the edge of the old roast yard, although, according to Hutchinson and Symington (1997), some metal stress still exists in the adjacent forest. They found that copper and nickel levels in surface soils up to 300 m from the roast bed were around ten or more times the background values (115–667 μg g^{-1} Ni and 149–1621 μg g^{-1} Cu), while pH values were between 3.6 and 5.7. Currently, the only woody species close to the roast bed are white birch and a dense cover of sheep laurel (*Kalmia angustifolia* L.). Further from the roast bed is an open white birch woodland with an understorey of wavy hair grass (*Deschampsia flexuosa* (L.) Trin.) and scattered spruces and pines.

2.2   Sulphur dioxide emissions, their effect on tree foliage and lichens, and indirect effects on plants

In 1945, Murray and Haddow reported 'severe burns' of tree foliage as far as 35 km to the north-east, 20 km to the north and 20 km to the south of the smelters. A lichen study indicated that only crustose lichens and rock-encrusting *Stereocaulon* sp. (probably *Stereocaulon saxatile* Magn. or *Stereocaulon paschale* (L.) Hoffm.) were to be found within the most highly polluted area, although *Parmelia physodes* (L.) Ach. (= *Hypogymnia physodes* (L.) Nyl.) and *Parmelia saxatilis* (L.) Ach. occurred near the edge of this zone. At a greater distance from the smelters, it was possible to find more sensitive lichen species, such as *Parmelia conspersa* (Ehrh. ex Ach.) Ach. and the caribou lichen (*Cladonia rangiferina* (L.) Wigg. = *Cladina rangiferina* (L.) Nyl.). LeBlanc and Rao (1966, 1973) found that lichens transplanted on to trees in the polluted zone did not survive for long, and in 1972, LeBlanc *et al.* used indices of atmospheric purity based on the occurrence of corticolous lichens on balsam poplar (*Populus balsamifera* L.) bark to prepare a map of pollution zones in Sudbury. They found a close correspondence with the direct atmospheric pollution measurements by Dreisinger (1965).

A similar pattern with respect to higher plants was noted by Linzon (1958), who noted the extreme sensitivity of white pine to sulphur dioxide in the area, and by Gorham and Gordon (1960a, b), who reported the absence of white pine and velvet-leaf blueberry (*Vaccinium myrtilloides* Michx.) within 24 km of the Falconbridge smelter, and a sharp drop in species numbers 6.4 km from the smelter.

Dreisinger and McGovern (1964) showed that under Sudbury field conditions, acute SO_2 injury to vegetation could occur when the following concentrations were equalled or exceeded:

- 0.95 ppm SO_2 for 1 h
- 0.55 ppm SO_2 for 2 h
- 0.35 ppm SO_2 for 4 h
- 0.25 ppm SO_2 for 8 h.

If any one of these conditions was met in daylight hours from mid-May to mid-October, then the fumigation was termed a 'Potentially Injurious Fumigation' (PIF). In a later Ministry of the Environment report, Dreisinger and McGovern (1969) drew pollution zones on a map of the Sudbury area, based on mean levels of SO_2 from 1953 to 1967 (Fig. 7.5). The most severely impacted zone formed an ellipse running from south-west to north-east of the Falconbridge smelter, surrounded by a less severe elliptical zone that encompassed the Coniston and Copper Cliff smelters. Dreisinger (1970) reported that during the period 1964–1968, an area of 5600 km^2 was impacted by SO_2 levels averaging 0.005 ppm, while 630 km^2 experienced mean levels of 0.02 ppm or more.

Estimates of the peak period for sulphur dioxide emission by Inco Ltd vary from the early 1960s to the early 1970s (Winterhalder, 1996). In recent years, improvements in the milling and flotation processes have led to the more complete rejection of the high-sulphur iron ore (pyrrhotite), and the recovery of the sulphur dioxide as marketable products such as liquid sulphur dioxide and sulphuric acid. The proportions of the rejected pyrrhotite being roasted for sulphuric acid production, stockpiled or mixed into the tailings stream has varied, with the latter strategy being the one most recently favoured. The development of sulphuric acid production by Inco began in the 1930s, and has culminated in the adaptation of its oxygen flash-smelting furnace for copper to the processing of a combined nickel–copper concentrate. By 1994, using these combined strategies, Inco reduced its sulphur dioxide emissions by 90% from the peak 1960 value to meet government limits of 265,000 tonnes per annum overall emission, as well as satisfying a 0.5 ppm hourly ground level average for stack emissions, and a 0.3 ppm half-hourly ground level average on fugitive

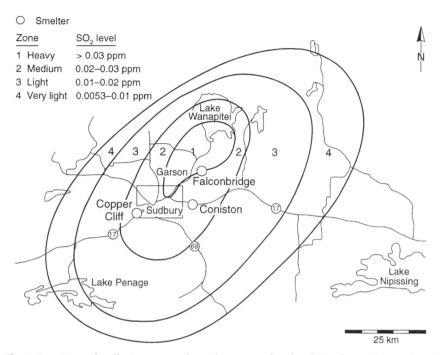

○ Smelter

Zone	SO₂ level
1 Heavy	> 0.03 ppm
2 Medium	0.02–0.03 ppm
3 Light	0.01–0.02 ppm
4 Very light	0.0053–0.01 ppm

Fig. 7.5.   Map of pollution zones, based on mean levels of SO_2 from 1953 to 1967 (adapted from Dreisinger and McGovern, 1969).

emissions from the smelter complex. Only 10% of the ore's sulphur content is currently released into the atmosphere.

In addition to monitoring air quality in isolation, the Ontario Ministry of the Environment published a series of reports on direct SO_2 damage to such species as poplars and birches, as well as cultivated plants (Dreisinger and McGovern, 1969; McGovern and Balsillie, 1972, 1973, 1974, 1975; McIlveen and Balsillie, 1978; Ontario Ministry of the Environment, 1982), often at great distance from the smelter. For example, in 1968, trembling aspen was damaged 110 km to the east of Sudbury, and 77 km to the south-west. In 1974, Hutchinson and Whitby (1974b) estimated that vegetation damage covered more than 5180 km².

Results of an experimental transplant of potted white pine in areas of high SO_2 pollution in 1969 did not support the hypothesis that high SO_2 fumigation was the primary factor limiting tree growth (Winterhalder, 1996). During intense fumigations, burns were often seen on leaves, but the effects were not fatal. It was clear, however, that long-term fumigation led to a decrease in soil pH. The indirect effects of low soil pH on plants are often more deleterious than the direct effects, and in a highly acidic soil that contains alumino-silicate minerals such as clays, aluminium ions will be present at a phytotoxic concentration, as will much of the copper and nickel from atmospheric fallout.

2.3   Metal deposition

During the late 1970s, Sudbury contributed only 40% of the total copper and nickel in wet deposition within 50 km of the Inco smelter (Chan *et al.*, 1982). Nevertheless, McIlveen (1990), in a study that plotted isopleths of surface concentrations of copper and nickel in 300 undisturbed soils across southern and north-eastern Ontario between 1980 and 1984, showed that the Sudbury complex dominated the distribution pattern of these two elements.

2.4   Soil erosion and bedrock weathering

The soils on the slopes in the area have a stony covering that is the result of the combined action of frost-heaving and erosion of the glacial till-derived soil. Extensive rock outcrops with shallow pockets of soil also occur. Lighter coloured patches on the blackened rocks show that physical weathering through exfoliation is in progress, although according to Pearce (1976), the rate of bedrock weathering in the Sudbury area ($50–170$ m^3 km^{-2} $year^{-1}$) is of the same order of magnitude as that for vegetated areas of similar climate and relief ($20–65$ m^3 km^{-2} $year^{-1}$). In contrast, the rate of soil erosion near Coniston (6000 m^3 km^{-2} $year^{-1}$) is two orders of magnitude higher than in comparable vegetated areas.

2.5   Frost action

Because of the paucity of vegetation and consequent lack of leaf litter, the soil surface lacks insulation and suffers from such cryogenic phenomena as needle ice and frost boil formation. This can lead to breakage of roots, or seedlings being heaved out of the soil.

2.6   Factor interaction

The many environmental factors impinging upon the Sudbury ecosystem do not act independently, but in an interactive and holistic manner, and their interactions often involve positive or negative feedbacks. While it is not possible to show all of these interactions schematically, an attempt has been made (Fig. 7.6) to show that a number of different factors are responsible for the Sudbury barrens.

3   The zonation of vegetation resulting from smelter impact

Extensive barren lands covering 17,000 ha form concentric zones around the three smelters, in turn surrounded by 72,000 ha of semi-barren birch

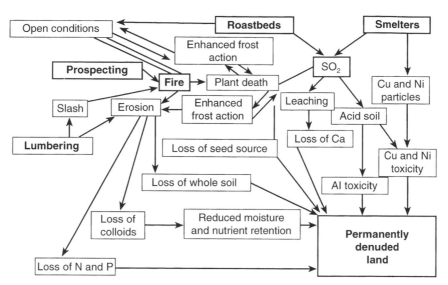

Fig. 7.6. A simplified representation of some of the major factor interactions leading to the creation and maintenance of the Sudbury barrens.

woodland (Fig. 7.7). This map, prepared by Struik (1973) from air photographs, bears a striking resemblance to the Dreisinger and McGovern (1969) map of pollution zones (Fig. 7.5). Pitblado and Amiro (1982) used remote sensing (LANDSAT digital data) to derive a vegetation index of relative biomass from the red and near infrared ratio, and showed a pattern similar to Struik's.

Amiro and Courtin (1981) classified the vegetation of the smelter-affected portion of the Sudbury area into nine major plant communities based on the composition of the tree strata. They concluded that three of these communities were unique, and directly attributable to the long-term effects of pollution. The first unique community, characterized by a soil pH of 4.0 and below, corresponded to the 'barren' zone of Fig. 7.7, and was essentially devoid of trees. The second and third communities, the 'Birch Transition' dominated by widely-spaced white birch, and the 'Maple Transition' dominated by red maple (*Acer rubrum* L.), accompanied by large-toothed aspen (*Populus grandidentata* Michx.), red oak (*Quercus borealis* Michx. f. = *Quercus rubra* L.)) and white birch, represented the 'semi-barren' woodland of Fig. 7.7.

3.1   The 'barrens'

Until recently, the only plants found on the 'barren' sites were highly depauperate relict individuals of woody species that had survived the stresses

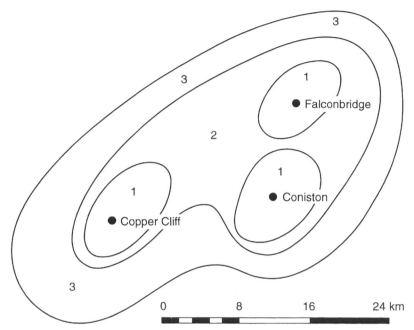

Fig. 7.7. Zones of vegetation damage, based on aerial photographs taken in 1970 (Struik, 1973). Zone 1 sites were barren, zone 2 sites semi-barren and zone 3 sites 'normal'.

responsible for the loss of the original vegetation. Tree species falling into this category are mostly white birch, red maple and red oak on drier sites, and trembling aspen in moister areas. While the first three species take on a coppiced form, with suckers arising from the base of a stool, the aspens exist as clonal patches arising from root suckers.

The relict red maples show a phenomenon best described as 'regressive dieback', in which the foliage reddens prematurely, and the amount of living biomass produced each year gradually decreases, the plant being surrounded by dead limbs of various sizes (Fig. 7.8). Relict white birch, less common than maple, shows premature yellowing of the margins of spur shoot leaves in early summer. However, later in the season, additional long shoot leaves are produced, which are totally green, and remain so until autumnal senescence.

Relict shrubs include the common blueberry, wild raisin (*Viburnum cassinoides* L.) and red elderberry (*Sambucus pubens* Michx. = *Sambucus racemosa* L.). The moss *Pohlia nutans* (Hedw.) Lindb. can be found in seepage areas on north-facing slopes, and the two pollution-tolerant lichens *Stereocaulon paschale* (L.) Hoffm. and *Cladonia deformis* (L.) Hoffm. (Nieboer *et al.*, 1972) sometimes colonize senescent moss patches or pockets of residual organic matter on blackened rock outcrops. In the Sudbury area, the

Fig. 7.8. Semi-barren 'birch transition' forest, with white birch and red maple. The former exhibits marginal chlorosis, while the latter (centre) is undergoing regressive dieback.

cryptogamic mat that is often characteristic of barren soils is typically composed of moss protonemata (especially those of *Pohlia nutans*), rather than the more usual filamentous algae.

More recently, metal-tolerant species have begun to colonize the barrens. Nevertheless, at the time of writing, there are still extensive barren areas that support no plant growth, or where growth is limited to metal-tolerant species.

3.2   The 'semi-barren woodlands'

The sparse understorey of the semi-barren communities consists of acid- and fire-tolerant species such as blueberry, sweet fern (*Comptonia peregrina* (L.) Coult. = *Myrica asplenifolia* Ait.) and bracken (*Pteridium aquilinum* (L.) Kuhn.), alternating with extensive bare areas. In damp depressions, the hair moss (*Polytrichum commune* Hedw.) forms dense carpets. The birch rarely reaches a height of more than 6 m and a diameter of 12 cm, and few individual stems exceed 30 years of age (James and Courtin, 1985; Trépanier, 1985). It is generally agreed that these stunted and depauperate woodlands are relict stands, as is evident from the predominantly coppiced form of the component tree species. According to Courtin (1995), stools from which the individual coppices arise are considerably older than those in traditional European coppice woodlands, but attempts to age the stools have been unsuccessful as a result of the chaotic ring pattern resulting from the cyclic regrowth of suckers.

4   Soil pH and metal content

Soil characteristics show a pattern of decreasing pH and increasing copper and nickel content towards the three smelters. When soil samples were taken in 1969 from each of the Dreisinger and McGovern (1969) pollution zones, chemical analysis showed a mean pH of around 4.0 in the 'heavy' zone, around 4.5 in the 'medium' zone and around 4.8 in the 'light' zone. The pH of mineral soils collected from hilltops and slopes in a south–south-east transect from the Copper Cliff smelter in the early 1970s ranged from 3.3 at 2 km distance to 3.9 at 34 km (Freedman and Hutchinson, 1980a), and Sudbury area soils were not only found to be acidified from the normal pH of 4.5–5.5 to pH 3.2–4.4, but the titratable acidity was found to be very high (2.5–6 meq 100 g^{-1}) and the base saturation from 1 to 5%. Extremes of soil acidity that have been recorded near the Coniston smelter include a pH of 2.4 (Hazlett *et al.*, 1983) and a pH of 2.2 (Hutchinson and Whitby, 1974a). Within any broad pH zone, there is inevitably a pattern of variability resulting from differential parent material, clay and/or humus content, degree of erosion, etc., as shown by Pitblado and Gallie (1995) using simple geographical information system (GIS) technology.

Soil copper and nickel levels also showed the expected trend in relation to fumigation zones (Fig. 7.9), and Taylor and Crowder (1983) showed a similar trend in wetland soils. The effects of smelters on the metal levels in soils are well known (Hutchinson, 1979). In most 'normal' soils, the average total copper content ranges from 15 to 40 $\mu g\ g^{-1}$ (average = 20 $\mu g\ g^{-1}$) (Aubert and Pinta, 1977), whereas Hazlett *et al.* (1983) found a copper content of 9700 $\mu g\ g^{-1}$ in

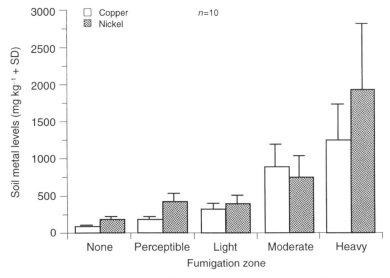

Fig. 7.9.   Increase in soil metal levels as the smelters are approached.

a soil 0.4 km from the Coniston smelter, while even 3 km from the Copper Cliff smelter, Freedman and Hutchinson (1980a) described a soil containing 3700 µg g^{-1} copper in its organic horizon. Phytotoxic symptoms often appear at 'available' copper levels between 25 and 50 µg g^{-1}, the symptoms becoming more severe as the pH drops, but Roshon (1988) found Sudbury-area soils with available copper ranging from 300 to 900 µg g^{-1} in her study.

Nickel contamination is also widespread in Sudbury soils. In most temperate and boreal soils, the total nickel content ranges from 20 to 30 µg g^{-1}, whereas Hazlett et al. (1983) found a nickel content of 6960 µg g^{-1} in a soil 0.2 km from the Coniston smelter. At a site 3 km from the Copper Cliff smelter, Freedman and Hutchinson (1980a) described a soil containing 3000 µg nickel g^{-1} in its surface horizon. The Ontario Ministry of the Environment also published a series of reports showing elevated metal levels in both soil and vegetation (McGovern and Balsillie, 1973, 1975; McIlveen and Balsillie, 1978; Ontario Ministry of the Environment, 1982). In a study south-east of the Coniston smelter, Rutherford and Bray (1979) found the highest total soil copper and nickel concentrations in depressions in summit areas and in valley basins. Furthermore, poorly drained soils showed a higher level of available (pH 3 NH$_4$OAc extractable) copper and nickel.

More recent studies (Gundermann and Hutchinson, 1993; Dudka et al., 1995; Negusanti, 1995; Sargent, 1996) have shown that a decline in copper and nickel levels is in progress in Sudbury area soils, as well as in the bioavailability and phytotoxicity (Gundermann and Hutchinson, 1993; Sargent, 1996). Possible mechanisms of reduction include erosion, acid leaching (Hutchinson, 1980) and increased rate of organic matter mineralization as soil pH is ameliorated.

5   Differential stress reaction of individual species

5.1   Red maple

The 'regressive dieback' described earlier is characteristic of red maples, both in the barren and semi-barren zone (Fig. 7.8). Coppiced individuals undergo progressive 'regressive dieback', in which less and less leaf biomass is produced each year. Each of the stems surrounding the base or stool eventually dies, and is replaced by a smaller stem. Ultimately all meristematic sites for suckering are used up, and the individual dies. Individuals continue to produce seeds throughout this prolonged period of dieback, and the seeds germinate normally, but quickly develop a red coloration and do not progress beyond the first foliar-leaf stage, even near the inactive smelter where birch establishment is common. The interaction of the red maple seedlings with the soil is clearly very different from that which occurs with white birch. There is some preliminary evidence that magnesium deficiency (or magnesium–calcium imbalance) may be a factor in the decline of red maple, but other possible causes, such as lack of vesicular-arbuscular (V-A) mycorrhizal infection, are also receiving attention.

The most striking feature of Sinclair's (1996) study of the vegetation of the Sudbury area is the disappearance of the red maple-dominated 'maple transition' community (Amiro and Courtin, 1981) over a period of 15 years, as a result of the regressive dieback.

Despite the apparent sensitivity of red maple to the environment of the Sudbury barrens, Watmough and Hutchinson (1997) have shown a signifi-cant positive linear relationship between the nickel tolerance of red maple callus tissue and the total extractable nickel level of the soil at the base of the tree of origin. They suggest that red maple is able to produce an adaptive response to metal exposure that could be crucial for its survival in a rapidly changing environment. Nevertheless, as mature trees continue to undergo regressive dieback, and seedlings of red maple are unable to survive more than a few months on metal-contaminated soils, this adaptive response is not an effective protection mechanism.

5.2   Red oak

Relict red oak can be found on the barrens, showing some of the dieback found in red maple, but the damage is not so severe, and does not appear to be regressive. Red oaks forming part of the semi-barren communities on ridges or on well-drained gravelly soils also show some dieback, often due to drought or to fire. Healthy red oak seedlings are found within oak stands, but not on the barrens, perhaps due to lack of seed dispersal to the extensive barren areas, as a result of their large seed size and specialized mode of dispersal by rodents such as squirrels.

5.3   Trembling aspen

The survival of trembling aspen as a relict is probably largely due to its ability to form root suckers, the trees usually occurring as clonal patches on lowland sites, rather than as individuals. They do not display conspicuous foliar symp-toms of stress. They produce copious seeds, but the seeds do not germinate unless the soil is limed.

5.4   White birch

White birch occurs as coppiced individuals that show a mid-season marginal chlorosis described above. Trees found on shallower soils and on more exposed ridge sites are more severely affected by the syndrome. Chlorotic portions of the leaf contain higher concentrations of copper and nickel and lower concentrations of calcium and magnesium than green portions. McIlveen and Negusanti (1984) showed that liming the soil alleviated the symptoms, especially by the second year after liming. Copious seeds are produced, some of which germinate and become established, especially near the inactive Coniston smelter.

5.5   Red pine

Occasional individuals of apparently relict red pine can be found scattered in semi-barren communities, where they usually appear to be quite healthy.

5.6   White pine

Unlike red pine, sulphur dioxide-sensitive white pine is not found in semi-barren communities. Nevertheless, following the improvement in atmospheric purity, occasional young trees have been found in the barrens. These young trees are always associated with the base of rock outcrops or boulders, where they are rooted in newly formed skeletal soil formed by the physical weathering of rock, rather than the metal-contaminated old soil.

6   Uptake and cycling of metals

Metal levels in plants show the expected relationship to distance from smelters. There is evidence of movement of metals through the food chain, although neither copper nor nickel bioaccumulates, and there is no evidence of metals reaching harmful levels in animals. A more detailed account of research in this area can be found in Winterhalder (1996).

7   Natural recovery – vegetation and soil dynamics following the reduction in atmospheric pollution

In 1972, the Inco smelter at Coniston and the pyrrhotite sintering plant at Falconbridge were closed, SO_2 output from the Inco iron ore recovery plant was limited to 250 tons (227.5 tonnes) day^{-1}, and the 381 metre 'superstack' at the Inco Copper Cliff smelter began operation. Consequently, the SO_2 tonnages emitted by both companies were significantly reduced. Direct evidence for the increase in atmospheric purity can be seen in the Ministry of the Environment's observations of reduced PIF and ground-level sulphur dioxide concentrations (McIlveen and Balsillie, 1978).

In 1978 and 1989–90, Beckett (1995) repeated observations on lichens growing on the bark of balsam poplar in the Sudbury area that had been made in 1968 (LeBlanc *et al.*, 1972). He found that the 'lichen desert' around the smelters had disappeared in the 20-year period, and that there was increased lichen richness and percentage cover in the outer lichen zones. Lichens are also invading rocks and soils, but more slowly. The caribou lichens, *Cladina rangiferina* and *Cladina mitis*, which are characteristic boreal forest understorey species, have changed little in distribution since 1977, occurring 20–25 km from the smelters (Cox, 1993).

Natural revegetation of the barrens has been slow, and limited to specific sites and mainly metal-tolerant species, confirming the hypothesis that soil properties are a more critical factor in the suppression of colonization than air quality. Immediate recovery in the barren zone was at first confined to moist, sheltered, nutrient-enriched sites, such as stream channels. In more exposed barren areas, minimal colonization by metal-tolerant tickle grass (*Agrostis scabra* Willd.) and tufted hairgrass (*Deschampsia cespitosa* (L.) P. Beauv.) did not occur until several years after the initiation of atmospheric improvement. On the semi-barren sites, improved growth of previously stunted birches and poplars has occurred since 1972. All species except red maple either maintained or increased their size and vigour in the 20 years following 1970, whereas the red maple, both on the barrens and in the woodland, continued to undergo regressive dieback.

Courtin (1995) has suggested that the slowness with which vegetation cover is spreading on to the barren areas between the birches of the semi-barren communities is due to a microclimate effect. While frost action is extreme in the barren areas, the area under each tree forms an 'oasis' by virtue of the insulating effects of the leaf litter. He predicts that these severe surface microclimatic conditions will only improve when the development of the ground and/or tree layer provides: (i) a closed canopy to reduce radiation and re-radiation; (ii) a litter layer that is not redistributed by the wind; and (iii) a network of roots to bind the soil.

7.1   Dynamics of the barren and semi-barren communities – species level changes

Colonization by plants showing genetically based tolerance

In the early 1970s, certain grass species began to colonize otherwise barren, metal-contaminated soils, the most notable being tufted hairgrass (Fig. 7.10), which has been shown to be multiple-metal-tolerant (Cox and Hutchinson, 1980). In 1995, Archambault and Winterhalder (1995) demonstrated that the Sudbury population of tickle grass, which had also been observed on the barren soils as long ago as 1972, possessed enhanced metal tolerance.

Two introduced grass species that have colonized barren sites to a more modest degree are redtop (*Agrostis gigantea* Roth, referred to by Freedman and

Fig. 7.10.   Metal-tolerant tufted hairgrass colonizing barren, phytotoxic soil.

Hutchinson (1980b) as *Agrostis stolonifera* var. *major*), and Canada bluegrass (*Poa compressa* L.). Although commercially available seed of both of these species is currently used in the revegetation operations, the populations seen colonizing barren land are often distant from revegetation sites, and it is likely that they have arisen from populations that pre-date revegetation activities. Enhanced metal tolerance has been demonstrated in both of these grasses. Hogan *et al.* (1977) found copper-tolerant strains of redtop on the copper- and nickel-rich surface of the old O'Donnell roast bed, some of which were later found (Hogan and Rauser, 1979) to be nickel tolerant, while Rauser and Winterhalder (1985) found enhanced zinc tolerance in some Canada blue-grass individuals near the old Coniston roast bed.

The shrub species dwarf or bog birch (*Betula pumila* L. var. *glandulifera* Regel) has shown a surprising ability to colonize barren land. It began to move on to barren, stony slopes from a small fen in the early 1980s (Fig. 7.11). Roshon (1988) has shown that there has been some genetic selection for metal tolerance in the Sudbury population of dwarf birch, but it is suspected that its success is at least partly due to lack of competition and the enhanced moisture supply provided by runoff from the numerous rock outcrops.

Colonization by plants apparently lacking genetically based tolerance

Near the Coniston smelter, closed since 1972, white birch, a typical boreal forest pioneer, began to colonize vigorously in the mid-1980s. In the absence of smelter emissions since 1972, soil pH changes of up to one unit have been

Fig. 7.11.   Dwarf birch moving on to barren, stony slopes from a small fen.

observed in the vicinity of the Coniston smelter over the past 20 years (Sargent, 1996). So far there is no evidence that white birch possesses metal-tolerant ecotypes, and it is likely that its ability to colonize these moderately toxic soils is the result of the phenotypic plasticity of the species. In contrast, seedlings of trembling aspen are never found on these barren soils. Although the predominant cause of the pH change was probably the leaching of free acids, it is possible that weathering of residual glacial till material released bases such as calcium and potassium, which displaced some of the adsorbed hydrogen ions, which were in turn lost through leaching.

Another Coniston area colonist that may not involve genetic selection for metal tolerance is wavy hairgrass. This grass is a conspicuous understorey species in birch- or oak-dominated semi-barren woodland, but is beginning to appear in previously barren sites that have been colonized by white birch. It is probably the species' well-known tolerance of low pH (Scurfield, 1954; Larcher, 1975) and very high soluble aluminium levels (Howson, 1984), rather than tolerance of metals, that allowed it to follow the white birch, its normal associate, on to the barrens.

7.2   Dynamics of the barren and semi-barren communities – community level changes

Sinclair (1996) has documented some of the changes in vegetation that have occurred in Amiro and Courtin's (1981) barren and semi-barren communities

('Barren', 'Birch Transition' and 'Maple Transition') since their study. Using similar clustering and ordination techniques to Amiro and Courtin's, she found that the maple transition community had disappeared over a period of 15 years as a result of regressive dieback.

8   Revegetation research – overcoming the limiting factors

Apart from early initiatives involving the use of imported soil, no attempt was made to revegetate the barrens until 1969, because it was generally believed that the only factor directly limiting the re-establishment and growth of plants was atmospheric quality. Therefore, the announcement by Inco Ltd in 1969 of its intention to construct a 381 m smoke-stack stimulated the Ontario Department of Lands and Forests (Sudbury District, Timber Branch) and Laurentian University Biology Department to initiate a joint experimental tree-planting programme on a barren site near Coniston and in a stunted, partially barren red oak woodland near Skead.

Survival of coniferous species in the semi-barren zone was good, although growth was slow, but in the barren sites mortality was close to 100%. It was soon realized that the soil was phytotoxic. Soil bioassay trials show restriction of seed germination and inhibition of root growth to a degree closely related to proximity to the smelters. Roots of seedlings germinated in soil from the barrens show an immediate cessation in growth when the soil is contacted, and the seedling develops a mass of stubby roots on top of the soil. Once the soil dries out, the seedling dies of drought. Liming of the soil reduces soil toxicity markedly, allowing the roots to penetrate the soil.

8.1   Limestone application as a soil detoxifier and trigger factor

The simple act of applying ground limestone to these toxic soils creates positive feedback loops throughout the system, and has an immediate detoxifying effect, both in the greenhouse and field situation. Indeed, surface application of limestone to toxic, barren soils triggers immediate colonization by native plant species, through germination both of the existing small seed bank and the incoming wind-disseminated seed rain. This minimal amelioration is not only more economical than the traditional approach, but it is a more effective means of engendering native plant colonization.

Although the primary factor limiting plant growth was clearly low pH combined with elevated copper and nickel values, later studies (Winterhalder, 1996) showed that aluminium also reached toxic levels.

8.2   The detoxification mechanism – neutralization and the roles of calcium and magnesium

The mechanism by which ground dolomitic limestone detoxifies Sudbury soils is not clear. When a contaminated soil is fully neutralized, a toxic metal can be precipitated as a carbonate or a hydroxide. In Sudbury soils, however, it is not necessary to raise the pH of a soil to more than around pH 5 to achieve detoxification. The hydroxylation of the trivalent aluminium ion as the pH is increased by liming is almost certainly a factor, but it does not fully explain detoxification, since theoretically the copper ions could still be active inhibitors of root growth. A possible role of both calcium and magnesium ions in the dolomitic limestone is in the competitive exclusion of metal ions from the root-hair's exchange complex. The calcium may also play a role in improving membrane integrity. McHale and Winterhalder (1997) have shown that in highly toxic soils with low liming rates, dolomitic limestone is a more effective detoxifier than calcitic limestone. It is hypothesized that this differential effect is the result of induced magnesium deficiency when the calcium : magnesium balance is altered by calcite application, and/or of antagonism between magnesium and nickel under dolomitic limestone treatment.

8.3   Soil phosphorus as a potentially limiting factor

Early field experiments, carried out on the sandy flood plain of Coniston Creek, indicated that soil phosphorus deficiency became a secondary limiting factor to plant growth once the soil was detoxified with limestone. However, it was later found that plant growth could be initiated and maintained for several years on stony slopes without the addition of phosphorus fertilizer, and it was hypothesized that there was a substantial capital of phosphorus, presumably in the form of phytin (inositol hexaphosphate) and its derivatives, in the 'fossil' organic matter derived from the pre-denudation vegetation. Most plants growing on these barren soils possess V-A mycorrhizae, assisting them in obtaining what little soil phosphorus is present.

9   Revegetation trials on steep, stony slopes

The loamy soils of the barren slopes, with their stony mantle, form an excellent seed bed and ameliorant trap, the stones forming a protective mulch. It was found in 1974 that a sprinkling of limestone, fertilizer and a grass–legume seed mixture on the soil surface in late summer gave rise to an effective vegetation cover (Fig. 7.12). Although frost heaving and needle ice formation were common in these soils, the enhanced root growth that followed liming allows for the stabilization of the heaved areas by plant roots, while the cracks formed

Fig. 7.12.   Experimental plots in which pulverized limestone, fertilizer and a grass seed mixture have been sprinkled on to the surface of barren, phytotoxic soil.

by frost action were beneficial in creating 'safe sites' in which seeds and ameliorants could lodge. In 1975 and 1977, 0.5 ha plots close to elementary schools were successfully revegetated in this manner by their students.

10   Revegetation – the operational procedures

In 1978, in response to an economic recession, the Regional Municipality of Sudbury acquired federal and provincial job-creation funding, as well as industry support, to employ 174 students who carried out several reclamation-related projects, including manual liming, fertilizing and seeding. The programme has continued on an annual basis, reaching its peak in 1983, when 1277 unemployed persons were employed on combined federal, provincial, municipal and industry funding in the summer months, working on Sudbury area revegetation. One of the major goals of the programme was image improvement, and the key visual corridors along major highways and urban neighbourhoods were targeted (Lautenbach, 1987a, b).

10.1   Establishment of a herbaceous cover

Pulverized dolomitic limestone was spread manually to give an application rate of approximately 10 tonnes ha^{-1}. After a few weeks to allow the limestone to react with the soil, 6-24-24 fertilizer was hand-spread at approximately

390 kg ha^{-1}, then seeding was carried out at a rate of 30–45 kg ha^{-1} in late August and September, using a manual cyclone seeder. The following is a typical seed mix (percentage composition by weight):

Agrostis gigantea Roth	Redtop	20%
Festuca rubra L.	Creeping red fescue	10%
Phleum pratense L.	Timothy	20%
Poa compressa L.	Canada bluegrass	15%
Poa pratensis L.	Kentucky bluegrass	15%
Lotus corniculatus L.	Birdsfoot trefoil	10%
Trifolium hybridum L.	Alsike clover	10%

Between 1978 and 1993, 3070 ha of barren land were grassed in this way, at a cost of approximately CAN$15 million (Lautenbach *et al.*, 1995). Inco Ltd has used a similar technique on its own barren land since 1980, employing all-terrain vehicles and, more recently, a commercial crop-dusting company to apply limestone, fertilizer and seeds from the air.

10.2   Tree planting

Spontaneous colonization of the grassed land by larger woody plants occurs almost immediately (Figs 7.13 and 7.14), but is so far confined to birch, poplar and willows. A natural conifer seed source is limited, and suitable seed bed conditions are rare. Furthermore, the 'greening' of the landscape that is so

Fig. 7.13.   Workers preparing to spread pulverized dolomitic limestone on a barren landscape in 1983.

Fig. 7.14. The same landscape in 1989, showing colonization of grassed land by trembling aspen.

evident in the summer is lost in the winter, because of the predominantly deciduous nature of the colonizing woody species. To make the winter landscape more visually pleasing and to ensure a ready seed source when seed bed conditions become available, species such as red pine, jack pine and white pine are planted in groups to act as a nucleus for future natural ecosystem development. Nursery stock of certain native deciduous species is also planted, of which red oak is probably the most successful.

Certain exotic tree species such as European larch (*Larix decidua* Mill.) and the small leguminous black locust (*Robinia pseudoacacia* L.) are also planted, raising questions concerning the advisability of using exotics when the creation of a quasi-natural plant community is the stated aim. On difficult sites, such as the silty-clay 'badlands' where the direct seeding approach does not work, black locust demonstrates a surprising tolerance of poor soil and frost (Fig. 7.15). In its native habitat, it is a vigorous pioneer colonist in the early regeneration of cleared or disturbed hardwood forest in the southern Appalachians (Boring and Swank, 1984). While there is some apprehension in the Sudbury area that the species might become something of a 'weed', as it has done in parts of Europe, its tendency towards a growth decrease and possible mortality after 10–20 years (Boring and Swank, 1984) suggests that it might play a useful role in soil stabilization and soil humus and nitrogen build-up before giving way to native woody species.

At the time of writing, over 3 million trees have been planted. Of the three pine species planted, jack pine shows the best survival rate (78%), closely

Fig. 7.15. Black locust growing vigorously on a highly eroded silty-clay soil.

followed by red pine (76%) and white pine (72%). The same pattern is seen with respect to growth rate, as indicated by mean annual increase in height, that of jack pine being 41 cm, red pine 33 cm and white pine 12 cm. In terms of annual increase in height, black locust holds the record, with an annual increase of 50 cm, despite the winter kill suffered by its youngest extension growth.

11 Vegetation dynamics following liming

Within a year of liming, there is conspicuous colonization by woody species such as trembling aspen, white birch and willows, as well as native 'wildflower' species such as pearly everlasting (*Anaphalis margaritacea* (L.) C.B. Clarke). As shown in Fig. 7.16, there is a tendency for the importance of introduced species to decrease and of native species to increase with time after treatment. The pattern shown by nitrogen-fixing birdsfoot trefoil (*Lotus corniculatus* L.) is interesting, since it peaks, then falls off as the canopy develops.

Allum and Dreisinger (1986) used satellite-based remote sensing to show qualitative changes in vegetation in response to combined atmospheric improvement and revegetation procedures undertaken by Inco Ltd, and were able to record changes over as little as 3 years. In Sinclair's (1996) study of vegetation change over 15 years, the emergence of a 'Birch–Pine' community was entirely dependent on the pines that had been planted as part of the revegetation programme, and which had grown to tree stature.

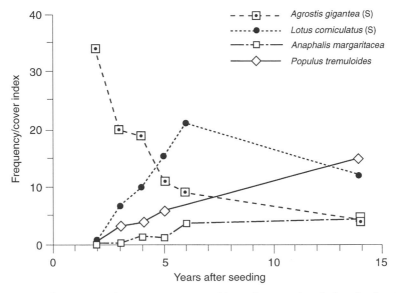

Fig. 7.16. Changes in relative importance of four plant species during the first 14 years following treatment. Species marked 'S' were seeded, while others were volunteers.

12   Native plant introduction by means of 'plug' transplants

One method of decreasing the limiting nature of the seed rain, and accelerating the establishment of a forest understorey, is to transplant blocks of soil and associated plants from natural communities, so that the plants can act as a seed source and a nucleus for vegetative spread. The use of plugs also introduces a small seed bank and source of native microbiota. A number of native species have been introduced in this way, including typical forest 'wildflowers' like Canada mayflower (*Maianthemum canadense* Desf.) and moccasin flower (*Cypripedium acaule* Ait.). The native microbiota that are introduced in this way, especially fungi and mites, are likely to have a beneficial effect on ecosystem function through enhanced leaf litter breakdown.

13   Soil dynamics following liming

13.1   pH and base dynamics on limed soil

A year or so after liming, the difference in pH between adjacent treated and untreated sites can be surprisingly low, often one pH unit or less, although successful revegetation has been achieved. At sites that were treated a number of years previously, larger differences, up two pH units, are sometimes found. It seems that an initial decline in pH following liming is followed by a rise, as bases are brought to the surface by the roots from the subsoil and deposited in leaf litter, a phenomenon referred to by Aber (1987) as a 'cation pump'. Aber points out, however, that some tree species are better cation pumps than others, with poplars and spruces being the best, pines actually acidifying the forest floor. The current procedure in the Sudbury area, of establishing a vigorous growth of birch and poplar before or concurrently with the planting of pines, therefore seems to be the correct choice.

13.2   Soil phosphorus dynamics

Beadle and Burges (1949) suggested that the size of the phosphorus 'working capital' in a soil or in a soil parent material can limit the type of plant community that will ultimately occupy the site. In the short term, phosphorus limitation in the Sudbury area has only been shown on sandy soils such as those of the Coniston Creek valley, where plot trials indicated that phosphorus deficiency was the secondary limiting factor following the pH–metal factor complex. Currently the Coniston Creek valley site, which was limed, fertilized and sown to Canada bluegrass in 1974, supports groves of trembling aspen, clonal patches of sweet fern and both jack pine and red pine plantations, so it may be that once plants are able to explore a deeper solum, the necessary phosphorus becomes available.

On stony slopes, experiments involving limestone-only treatments indicate that fertilization is not necessary in the short term, as the reservoir of phosphorus in the residual organic matter that was laid down prior to denudation is sufficient to support plant growth for several years, depending on the degree of erosion that has occurred. In addition, limestone application has a beneficial effect in making phosphorus in acidic soils available to plants. There is also a small reservoir of phosphorus in the unweathered glacial till pebbles that characterize these soils, and it seems that there is sufficient phosphorus in the system to support growth for a number of years. On the other hand, the sandy valley-bottom soils may remain very poorly vegetated until their phosphorus levels build up through accessions from blown-in detritus, visiting animals, occasional floods and airborne deposition.

13.3   Soil nitrogen dynamics

Nitrogen deficiency is considered to be one of the commonest limiting factors in land reclamation, since the primary source is atmospheric rather than terrestrial. The Sudbury barrens present much less of a nitrogen challenge than most reclamation sites because of the residual organic matter in the soil. Nevertheless, once the nitrogen in the soil is used up, a source will be required, and the use of biological nitrogen fixation must become part of the

revegetation formula if a maintenance-free system is the goal. Although potentially nitrogen-fixing cyanobacteria are absent from Sudbury's barren soils,
Maxwell (1991, 1995) observed that they colonized barren sites quickly after
liming, and she found evidence that the cyanobacteria inoculum was brought
on to the site on the limers' boots.

The mature native forest of the Sudbury area contains no leguminous species, and leguminous species play no role in succession following disturbance
in this area. The only nitrogen-fixer in upland sites is sweet fern, which plays a
role in jack pine succession. Relict stands of sweet fern occur on the barrens,
and often colonize revegetated sites and spread rapidly, especially on those
which have been limed but not seeded. However, the native alders (speckled
alder, *Alnus rugosa* (Du Roi) Spreng. = *Alnus incana* (L.) Moench. subsp. *rugosa*
(DuRoi) Clausen, and green alder, *Alnus crispa* (Ait.) Pursh.) do not readily
colonize treated land, despite their winged propagules. Research is underway
on the ways of enhancing sweet fern's contribution to the nitrogen cycle,
rather than relying on exotic species, although the role of sweet fern as the
alternate host of a fungal rust disease on jack pine is a possible problem.

14   Prognosis for permanent success

It is difficult to estimate the amount of time that a naturally recovering
contaminated site might take to reach a stable, quasi-natural community
structure and floristic composition, especially in view of the possibility of global
climate change. It seems likely, however, that the process will be much slower
than a similar phenomenon in an aquatic system, and that during the recovery process, soil and contaminants will continue to be eroded into the valleys,
the watercourses and finally the lakes. The anoxic sediments of these lakes
will ultimately constitute a major sink for toxic trace elements (Belzile and
Morris, 1995). It is therefore beneficial to accelerate plant establishment and
soil stabilization by revegetation procedures. Nevertheless, the liming of the
landscape brings about a change in the balance of elements in the system, the
effects of which may be difficult to predict.

Despite the slow release of phosphorus to the system by the weathering of
undecomposed stones in the glacial till and fragmented bedrock, edaphic
factors may prevent the development of the full climatic climax vegetation-
type. This is not necessarily a bad thing; subclimaxes of various sorts are found
all over the world, many of them the result of perfectly 'natural' factors. Bowler
(1992) is of the opinion that North Americans have too little appreciation of
the environmental values inherent in anthropogenic landscapes like the
maquis and garrigue of Mediterranean France, and he suggests that ecologically disturbed areas have a value as corridors and buffers, back-up habitat
for wildlife, multi-successional landscapes and as areas for the practice of
restoration.

In the case of semi-barren communities, James and Courtin (1985) and Courtin (1995) have suggested that poor seedling survival, periodic crown dieback, persistence of acidic, metal-contaminated soils, periodic drought and attack by phytophagous insects will maintain this woodland in its present coppiced form. Amiro and Courtin (1981) have questioned whether a full recovery of Sudbury's degraded sites can ever occur, because of the changes in soil chemistry, and Courtin (1995) suggests that even the open birch-coppice community will require the planting of conifers to increase tree species diversity sufficiently to escape the current cycle of insect infestations, which is characteristic of a monoculture.

A number of factors have shaped the Sudbury land reclamation programme, including government pollution control legislation during a prosperous period for the mining industry, effective partnerships between industry, government, academia and the public, a minimal treatment approach combined with a willingness to remain flexible and focus on achievable goals, and the direct involvement of the public (Gunn *et al.*, 1995). Land-use planning has been minimal, in spite of the fact that Pierpoint and Hills (1963) considered the Sudbury Basin to be 'one of the few places on the Canadian Shield where the land has a good potential for agricultural use and is of an area large enough to be suitable for a dominantly agricultural community'. Nevertheless, the current consensus is that the degraded landscape should be returned to something resembling its original forest cover.

References

Aber, J.D. (1987) Restored forests and the identification of critical factors in species-site interactions. In: Jordan, W.R. III, Gilpin, M.E. and Aber, J.D. (eds) *Restoration Ecology: a Synthetic Approach to Ecological Research*. Cambridge University Press, Cambridge, pp. 241–250.

Allum, J.A.E. and Dreisinger, B.R. (1986) Remote sensing of vegetation change near Inco's Sudbury mining complexes. *International Journal of Remote Sensing* 8, 399–416.

Amiro, B.D. and Courtin, G.M. (1981) Patterns of vegetation in the vicinity of an industrially disturbed ecosystem, Sudbury, Ontario. *Canadian Journal of Botany* 59, 1623–1639.

Archambault, D. J.-P. and Winterhalder, K. (1995) Metal tolerance in *Agrostis scabra* from the Sudbury, Ontario, area. *Canadian Journal of Botany* 73, 766–775.

Aubert, H. and Pinta, M. (1977) *Trace Elements in Soils*. Elsevier Scientific Publishing, New York.

Beadle, N.C.W. and Burges, A. (1949) Working capital in a plant community. *Australian Journal of Science* 11, 207–208.

Beckett, P.J. (1995) Lichens: sensitive indicators of improving air quality. In: Gunn, J. (ed.) *Environmental Restoration and Recovery of an Industrial Region*. Springer-Verlag, New York, pp. 81–91.

Belzile, N. and Morris, J.R. (1995) Lake sediments: sources or sinks of industrially mobilized elements? In: Gunn, J. (ed.) *Environmental Restoration and Recovery of an Industrial Region*. Springer-Verlag, New York, pp. 183–193.

Boring, L.R. and Swank, W.T. (1984) The role of black locust (*Robinia pseudoacacia*) in forest succession. *Journal of Ecology* 72, 749–766.

Bowler, P.A. (1992) Biodiversity conservation in Europe and North America II. Shrublands – in defense of disturbed land. *Restoration and Management Notes* 10, 144–149.

Chan, W.H., Ro, C., Lusis, M.A. and Vet, R.J. (1982) *Sudbury Environmental Study. An Analysis of the Impact of Inco Emissions on Precipitation Quality in the Sudbury Area*, Sudbury Environment Study Report SES 012/82. Ontario Ministry of the Environment, Toronto.

Courtin, G.M. (1995) Birch coppice woodlands near the Sudbury smelters: dynamics of a forest monoculture. In: Gunn, J. (ed.) *Environmental Restoration and Recovery of an Industrial Region*. Springer-Verlag, New York, pp. 233–245.

Cox, J.D. (1993) Survival strategies of lichens and bryophytes in the mining region of Sudbury. MSc thesis, Laurentian University, Sudbury, Ontario.

Cox, R.M. and Hutchinson, T.C. (1980) Multiple metal tolerances in the grass *Deschampsia cespitosa* (L.) Beauv. from the Sudbury smelting area. *New Phytologist* 84, 631–647.

Dreisinger, B.R. (1965) Sulphur dioxide levels and the effects of the gas on vegetation near Sudbury, Ontario. Paper 65–121. In: *Proceedings of the 58th Annual Meeting, Air Pollution Control Association, Toronto*.

Dreisinger, B.R. (1970) Sulphur dioxide levels and vegetation injury in the Sudbury area during the 1969 season. Ontario Department of Mines, Sudbury.

Dreisinger, B.R. and McGovern, P.C. (1964) Sulphur dioxide levels in the Sudbury area and some effects of the gas on vegetation in 1963. Ontario Department of Mines, Sudbury.

Dreisinger, B.R. and McGovern, P.C. (1969) Sulphur dioxide levels and resultant injury to vegetation in the Sudbury area during the 1968 season. Ontario Department of Mines, Sudbury.

Dudka, S., Ponce-Hernandez, R. and Hutchinson, T.C. (1995) Current level of total element concentrations in the surface layer of Sudbury's soils. *Science of the Total Environment* 162, 161–171.

Freedman, B. and Hutchinson, T.C. (1980a) Pollutant inputs from the atmosphere and accumulation in soils and vegetation near a copper–nickel smelter at Sudbury, Ontario. *Canadian Journal of Botany* 58, 108–132.

Freedman, B. and Hutchinson, T.C. (1980b) Long-term effects of smelter pollution at Sudbury, Ontario, on forest community composition. *Canadian Journal of Botany* 58, 2123–2140.

Gorham, E. and Gordon, A.G. (1960a) Some effects of smelter pollution north-east of Falconbridge, Ontario. *Canadian Journal of Botany* 38, 307–312.

Gorham, E. and Gordon, A.G. (1960b) The influence of smelter fumes upon the chemical composition of lake waters near Sudbury, Ontario, and upon the surrounding vegetation. *Canadian Journal of Botany* 38, 477–487.

Gundermann, D.G. and Hutchinson, T.C. (1993) Changes in soil chemistry 20 years after the closure of a nickel–copper smelter near Sudbury, Ontario, Canada.

In: Allan, R.J. and Nriagu, J.O. (eds) *Heavy Metals in the Environment*, Vol. 2. CEP Consultants, Edinburgh, pp. 559–562.

Gunn, J.M., Conroy, N., Lautenbach, W.E., Pearson, D.A.B., Puro, M.J., Shorthouse, J.D. and Wiseman, M.E. (1995) From restoration to sustainable ecosystems. In: Gunn, J. (ed.) *Restoration and Recovery of an Industrial Region*. Springer-Verlag, New York, pp. 335–344.

Hazlett, P.W., Rutherford, J.K. and vanLoon, G.W. (1983) Metal contaminants in surface soils and vegetation as a result of nickel/copper smelting at Coniston, Ontario, Canada. *Reclamation and Revegetation Research* 2, 123–137.

Hogan, G.D. and Rauser, W.E. (1979) Tolerance and toxicity of cobalt, copper, nickel and zinc in clones of *Agrostis gigantea*. *New Phytologist* 83, 665–670.

Hogan, G.D., Courtin, G.M. and Rauser, W.E. (1977) Copper tolerance in clones of *Agrostis gigantea* from a mine waste site. *Canadian Journal of Botany* 55, 1043–1050.

Howson, M. (1984) Comparative ecology of two *Deschampsia* species at Sudbury, Ontario, Canada. MSc thesis, Department of Botany, University of Toronto, Ontario.

Hutchinson, T.C. (1979) Copper contamination of ecosystems caused by smelter activities. In: Nriagu, J.O. (ed.) *Copper in the Environment Part 1. Ecological Cycling*. John Wiley & Sons, New York, pp. 451–502.

Hutchinson, T.C. (1980) Effects of acid leaching on cation loss from soils. In: Hutchinson, T.C. and Havas, M. (eds) *Effects of Acid Precipitation on Terrestrial Ecosystems*. Plenum Press, New York, pp. 481–497.

Hutchinson, T.C. and Symington, M.S. (1997) Persistence of metal stress in a forested ecosystem near Sudbury, 66 years after closure of the O'Donnell roast bed. *Journal of Geochemical Exploration* 58, 323–330.

Hutchinson, T.C. and Whitby, L.M. (1974a) Heavy-metal pollution in the Sudbury mining and smelting region of Canada. I. Soil and vegetation contamination by nickel, copper and other metals. *Environmental Conservation* 1, 123–132.

Hutchinson, T.C. and Whitby, L.M. (1974b) A study of airborne contamination on vegetation and soils by heavy metals from the Sudbury, Ontario copper–nickel smelters. In: Hemphill, D.D. (ed.) *Trace Elements in Environmental Health VII. A Symposium*. University of Missouri Press, Columbia, pp. 175–178.

James, G.I. and Courtin, G.M. (1985) Stand structure and growth form of the birch transition community in an industrially damaged ecosystem, Sudbury, Ontario. *Canadian Journal of Forest Research* 15, 809–817.

Larcher, W. (1975) *Physiological Plant Ecology*. Springer-Verlag, Heidelberg.

Lautenbach, W.E. (1987a) The Sudbury Regional Land Reclamation Program – a planner's perspective. In: Beckett, P.J. (ed.) *Proceedings of the 12th Annual Meeting, Canadian Land Reclamation Association, 7–11 June 1987, Sudbury, Ontario*. CLRA, Guelph, pp. 22–31.

Lautenbach, W.E. (1987b) The greening of Sudbury. *Journal of Soil and Water Conservation* 42, 228–231.

Lautenbach, W.E., Miller, J., Beckett, P., Negusanti, J. and Winterhalder, K. (1995) Municipal land restoration program: the regreening process. In: Gunn, J. (ed.) *Environmental Restoration and Recovery of an Industrial Region*. Springer-Verlag, New York, pp. 109–122.

LeBlanc, F. and Rao, D.N. (1966) Réaction de quelques lichens et mousses épiphytiques à l'anhydride sulfureux dans la région de Sudbury, Ontario. *Bryologist* 69, 338–346.

LeBlanc, F. and Rao, D.N. (1973) Effects of sulphur dioxide on lichen and moss transplants. *Ecology* 54, 612–617.

LeBlanc, F., Rao, D.N. and Comeau, G. (1972) The epiphytic vegetation of *Populus balsamifera* and its significance as an air pollution indicator in Sudbury, Ontario. *Canadian Journal of Botany* 50, 519–528.

Linzon, S.N. (1958) *The Influence of Smelter Fumes on the Growth of White Pine in the Sudbury Region*, Technical Report, Ontario Department of Lands and Forests, Department of Mines. Department of Lands and Forests, Toronto.

Maxwell, C.D. (1991) Floristic changes in soil algae and cyanobacteria in reclaimed metal-contaminated land at Sudbury, Canada. *Water, Air, and Soil Pollution* 60, 381–393.

Maxwell, C.D. (1995) Acidification and heavy metal contamination: implications for the soil biota of Sudbury. In: Gunn, J. (ed.) *Environmental Restoration and Recovery of an Industrial Region*. Springer-Verlag, New York, pp. 219–231.

McGovern, P.C. and Balsillie, D. (1972) *Sulphur Dioxide Levels and Environmental Studies in the Sudbury Area During 1971*. Technical Report, Ontario Ministry of the Environment, Air Quality Branch, Ontario Ministry of the Environment, Sudbury.

McGovern, P.C. and Balsillie, D. (1973) *Sulphur Dioxide (1972) – Heavy Metal (1971) Levels and Vegetative Effects in the Sudbury Area*. Ontario Ministry of the Environment, Air Management Branch. Ontario Ministry of the Environment, Sudbury.

McGovern, P.C. and Balsillie, D. (1974) *Effects of Sulphur Dioxide and Heavy Metals on Vegetation in the Sudbury Area (1973)*. Ontario Ministry of the Environment, Sudbury.

McGovern, P.C. and Balsillie, D. (1975) *Effects of Sulphur Dioxide and Heavy Metals on Vegetation in the Sudbury Area (1974)*. Ontario Ministry of the Environment, Northeast Region, Ontario Ministry of the Environment, Sudbury.

McHale, D. and Winterhalder, K. (1997) The importance of the calcium–magnesium ratio of the limestone used to detoxify and revegetate acidic, nickel- and copper-contaminated soils in the Sudbury, Canada mining and smelting area. In: Jaffré, T., Reeves, R.D. and Becquer, T. (eds) *The Ecology of Ultramafic and Metalliferous Areas. Proceedings of the Second International Conference on Serpentine Ecology, Nouméa, July 31–August 5, 1995*. ORSTOM, Paris, pp. 267–273.

McIlveen, W.D. (1990) Terrestrial effects study. In: Anon. (ed.) *Acidic Precipitation in Ontario Study (APIOS) Annual Program Report 1988/1989*. Ontario Ministry of the Environment, Toronto.

McIlveen, W.D. and Balsillie, D. (1978) *Air Quality Assessment Studies in the Sudbury Area. Vol. 2. Effects of Sulphur Dioxide and Heavy Metals on Vegetation and Soils 1970–1977*. Ontario Ministry of the Environment, Technical Support Section, Northeastern region, Ontario Ministry of the Environment, Sudbury.

McIlveen, W.D. and Negusanti, J.J. (1984) *Marginal Chlorosis of White Birch in the Sudbury Area, 1978–1983*. Ontario Ministry of the Environment Progress Report NER-AQTM 11-84, Ontario Ministry of the Environment, Sudbury.

Murray, R.H. and Haddow, W.R. (1945) First Report of the Sub-committee on the investigation of sulphur smoke conditions and alleged forest damage in the Sudbury Region, February 1945. Unpublished report.

Negusanti, J.J. (1995) Terrestrial metal trends in the Sudbury area. In: Hynes, T.P. and Blanchette, M.C. (eds) *Proceedings of Sudbury '95 – Mining and the Environment. 28 May–1 June 1995, Sudbury, Ontario, Canada*, 3 Vols. CANMET, Ottawa, pp. 1143–1150.

Nichols, G.E. (1935) The Hemlock–White Pine–Northern Hardwoods region of eastern North America. *Ecology* 16, 403–422.

Nieboer, E., Ahmed, H.M., Puckett, K.J. and Richardson, D.H.S. (1972) Heavy metal content of lichens in relation to distance from a nickel smelter in Sudbury, Ontario. *Lichenologist* 5, 292–303.

Ontario Ministry of the Environment (1982) *Sudbury Environmental Study Synopsis: 1973–1980*. Ontario Ministry of the Environment, Sudbury.

Pearce, A.J. (1976) Geomorphic and hydrologic consequences of vegetation destruction, Sudbury, Ontario. *Canadian Journal of Earth Sciences* 13, 188–193.

Pearson, D.A.B. and Pitblado, J.R. (1995) Geological and geographic setting. In: Gunn, J. (ed.) *Environmental Restoration and Recovery of an Industrial Region*. Springer-Verlag, New York, pp. 5–15.

Pierpoint, G. and Hills, G.A. (1963) *The Land Resources of the Sudbury Basin*. Research Report No. 50, Ontario Department of Lands and Forests, Toronto.

Pitblado, J.R. and Amiro, B.D. (1982) Landsat mapping of the industrially disturbed vegetation communities of Sudbury, Canada. *Canadian Journal of Remote Sensing* 8, 17–28.

Pitblado, J.R. and Gallie, E.A. (1995) Remote sensing and geographic information systems: technologies for mapping and monitoring environmental health. In: Gunn, J. (ed.) *Environmental Restoration and Recovery of an Industrial Region*. Springer-Verlag, New York, pp. 299–311.

Rauser, W.E. and Winterhalder, E.K. (1985) Evaluation of copper, nickel, and zinc tolerances in four grass species. *Canadian Journal of Botany* 63, 58–63.

Roshon, R. (1988) Genecological studies on two populations of *Betula pumila* var. *glandulifera*, with special reference to their ecology and metal tolerance. MSc thesis, Laurentian University, Sudbury, Ontario.

Rowe, J.S. (1959) *Forest Regions of Canada*, publication 1300, Department of the Environment, Canadian Forestry Service, Information Canada, Ottawa.

Rutherford, G.K. and Bray, C.R. (1979) Extent and distribution of soil heavy metal contamination near a nickel smelter at Coniston, Ontario. *Journal of Environmental Quality* 8, 219–222.

Sargent, D.G. (1996) The response of the terrestrial ecosystem at Sudbury, Ontario to 20 years of reduced smelter pollution. MSc thesis, Trent University, Peterborough, Ontario.

Scurfield, G. (1954) Biological flora of the British Isles. *Deschampsia flexuosa* (L.) Trin. *Journal of Ecology* 42, 225–233.

Sinclair, A. (1996) Floristics, structure and dynamics of plant communities on acid, metal-contaminated soils. MSc thesis, Laurentian University, Sudbury, Ontario.

Struik, H. (1973) Photo interpretive study to assess and evaluate the vegetational and physical state of the Sudbury area subject to industrial emissions. In: Anon. (ed.) *Sudbury Environmental Enhancement Programme Summary Report, 1969–1973*. Department of Lands and Forests, Ontario, Department of Lands and Forests, Sudbury, pp. 6–11.

Taylor, J.G. and Crowder, A.A. (1983) Accumulation of atmospherically deposited metals in wetland soils of Sudbury, Ontario. *Water, Air, and Soil Pollution* 19, 29–42.

Trépanier, M. (1985) Stem analysis of white birch in the birch transition community, Sudbury, Ontario. BSc thesis, Department of Biology, Laurentian University, Sudbury, Ontario.

Watmough, S.A. and Hutchinson, T.C. (1997) Metal resistance in red maple (*Acer rubrum*) callus cultures from mine and smelter sites in Canada. *Canadian Journal of Forest Research* 27, 693–700.

Winterhalder, K. (1996) Environmental degradation and rehabilitation of the landscape around Sudbury, a major mining and smelting area. *Environmental Reviews* 4, 185–224.

Impact of Air Pollution on the Forests of Central and Eastern Europe

8

K. Vancura,[1] G. Raben,[2] A. Gorzelak,[3] M. Mikulowski,[3] V. Caboun[4] and J. Oleksyn[5,6]

[1]Forestry and Game Management Research Institute (ULHM), Prague, Czech Republic; [2]Saxon State Institute for Forestry (LAF), Graupa, Germany; [3]Forest Research Institute (IBL), Warsaw, Poland; [4]Forestry Research Institute (LVU), Zvolen, Slovakia; [5]Polish Academy of Sciences, Institute of Dendrology, Kórnik, Poland; [6]University of Minnesota, Department of Forest Resources, St Paul, USA

Over the past 50 years, there has been an unprecedented deterioration of the environment of Central and Eastern European (CEE) countries. Air pollutants from fossil fuel combustion and heavy industry threaten productivity and in some places the future survival of forests. Political changes at the end of the 1980s were followed by economic transformations and significant declines in pollutant emissions. The decline in emissions was particularly noticeable (more than 50% in some countries) for sulphur dioxide. However, reductions in the atmospheric concentrations of toxic gases have not eliminated local and regional forest declines or the existence of pollution 'hot spots' with average daily concentrations of SO_2 up to 1000 µg m^{-3}. Changes in defoliation levels have not followed reductions in pollutant emissions and remain high throughout the region. Imbalanced nutrition is an important factor affecting tree health and reducing the resistance of forest ecosystems to combinations of other stresses. The reasons for the lack of a direct relationship between regional pollution loads and tree crown condition are uncertain and should be further investigated. A new type of intensive monitoring (UN-ECE Level II) should provide regional information about the ecological risks facing forest ecosystems, and should also lead to recommendations for forest management in areas impacted by air pollution. In Belarus, the Russian Federation and Ukraine, over 3.5 million hectares of forests were affected by the Chernobyl nuclear disaster, resulting in the withdrawal of the wood and secondary products from the market and up to ten times higher than normal costs of forest operations.

1 Introduction

The forest ecosystems of Central and Eastern Europe (CEE) are an important resource for the region, with substantial ecological and social value. In the post-war period, the area of forest increased by 32% in Poland, 13% in the Czech Republic and Slovakia, and 10% in Germany (Vasicek, 1997). However, some countries in the region have a long history of industrial development and pollution, with the first symptoms of forest decline being described 150 years ago (Godzik and Sienkiewicz, 1990). Damage started in the vicinity of important point-source polluters, and was caused by high SO_2 concentrations (Rauchschäden). However, it is only since the latter half of the 20th century that CEE countries have experienced an unprecedented deterioration in their environments, which has affected the ecological integrity of the forests. In the 1980s, when peak levels of pollution in CEE coincided with the fall of communism and the abolition of censorship, damage to trees by industrial pollution became a major scientific and political issue. No specific pollutants have been implicated as causing forest damage in the region, partly because the observed damage symptoms (such as leaf discoloration and necrosis or crown dieback) are non-specific and can be caused by a variety of stresses, including nutrient deficiency, pathogens, severe drought and temperature anomalies (Schulze, 1989; Kauppi *et al.*, 1992; Ellsworth and Oleksyn, 1997). The results of the crown transparency assessments led to the widespread belief that forests in CEE countries were in a process of progressive death caused by air pollution (e.g. Bialobok, 1989). However, subsequent surveys and investigations have failed to confirm that forests are dying or declining as a result of environmental pollution over large areas of European countries (Kandler and Innes, 1995). Cases of severe decline of forests throughout Europe actually cover a maximum of 8000 km^2, or less than 0.5% of the forest area (Kauppi *et al.*, 1992).

Physiological symptoms, foliar injury, decreases of productivity and tree mortality caused by air pollutants are common phenomena in CEE countries. However, they are usually restricted to different point-source polluters and/or urban and industrial agglomerations (e.g. Upper Silesia, the Krakow region in Poland, and north-western Bohemia and the northern Moravia regions in the Czech Republic). In specific regions of Central Europe, forest decline triggered by air pollution stress is considered to be one of the most important environmental threats facing the forests. For example in the area called the 'Black Triangle' where the Polish, German and Czech borders meet, regional forest declines have occurred as a result of the high concentrations of several pollutants (mostly SO_2 and NO_x) and their combination with other types of stress (Innes, 1993). Interactions between air pollutants and other stresses can take several forms, but are often synergistic. As a result of the complexity of the possible responses, many effects may not be immediately apparent, although the simultaneous occurrence of several stresses can predispose trees to decline (Gorzelak, 1995). In addition to acute injury caused by high levels of gaseous air pollutants, there is an ongoing process of nutrient leaching from forest soils, especially of base cations, which leads to further increases in nutritional deficiencies and decreased site productivity. These in turn may lead to the loss of forest stability and impairment of the protective functions of forests.

A number of reviews on the mechanisms of damage and environmental problems associated with pollution impacts on forest have been published recently (Taylor *et al.*, 1994; Kandler and Innes, 1995; Matyssek *et al.*, 1995). Here, we focus on the nature of the industrial pollution problems and the risks they pose to CEE forests, either alone or in combination with other abiotic and biotic factors. We also present a detailed description of the environmental aftermath of the Chernobyl catastrophe.

2   Threats to Central and Eastern European forests

2.1   Pollution threats

In recent decades, air pollution has become the most serious threat facing the forests of CEE. As a result of the control measures introduced since the beginning of the 1980s (see Chapter 12, this volume), deposition of acidity and sulphur has been reduced substantially, especially in Germany (and other parts of Western Europe). However, nitrogen deposition levels are still very high and may even be increasing. A similar pattern occurred in the area of the former German Democratic Republic (GDR) after the unification of Germany in 1989 (Raben *et al.*, 1998), although in this case acidic deposition remained high at some monitoring plots in the Saxon part of the Ore Mountains.

Energy production technologies represent the largest sources of pollutant emissions in CEE. Factors contributing to pollution problems in some CEE countries are the scarcity of environmentally clean energy sources and the use of fossil fuel and combustion technologies for energy production. Coal-burning power generation plants are responsible for 44% of SO_2 emissions in Slovakia, 56% in Poland, 58% in the Czech Republic and 63% in Hungary and Germany (Environment, 1997). In Bulgaria, Germany, Poland and the Ukraine, coal burning provides from 46 to 95% of the primary energy (Fig. 8.1). For example in Poland during the past decade, increased demand for electric energy was met by the increased burning of low-quality, high-sulphur, soft coal. For economic reasons, soft coal is normally burned close to its source. Therefore, it

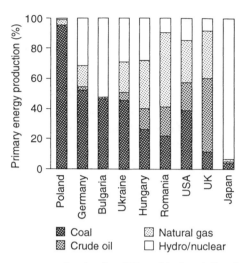

Fig. 8.1. Primary energy production by CEE and industrialized countries (Environment, 1997). Data are for coal, crude oil, natural gas liquids, natural gas, and hydroelectric and nuclear electricity.

often creates local environmental problems associated with SO_2, NO_x dust and volatile organic carbon compounds. In Poland soft-coal burning has increased by more than 80%, from 35 Mt in 1980 to 64 Mt in 1996 (Environment, 1997). Open-pit mining of soft coal also lowers water tables, adversely affecting local agricultural and forested areas. For example open-cast mines in northern Bohemia and Sokolov were excavated to a depth of 150–180 m, destroying over 26,000 ha of land, and leaving residual pits with a total volume of 3.5 million m^3 (Moldan, 1997). Mining activities in Poland have resulted in almost 34,000 ha of state forests being affected by drought due to the lowering of water tables (Forestry, 1997).

The current availabilty of natural resources suggests that the composition of energy sources and associated pollution will remain as a potential environmental threat for several decades. The combination of economic crisis, production restructuring and the introduction of more efficient technologies has lowered demands for electric energy in recent years. In comparison with 1980, primary energy consumption decreased in 1994 by 7% in unified Germany, 15% in Russia, 18% in Hungary, 35% in Romania, 36% in Poland and 43% in Bulgaria. Over the same period, primary energy consumption in western European countries increased by between 7 and 41%, by 13% in the USA and by 39% in Japan (Environment, 1997).

During the 1990s, there have been changes in the composition of air pollution in CEE. For example, in the Czech Republic, sulphur dioxide has significantly decreased, and hydrogen fluorides, ozone and acidic deposition are now considered to be the most important problems in the country (Fig. 8.2).

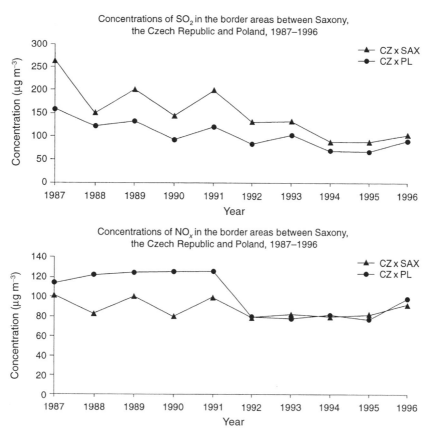

Fig. 8.2. Yearly trends in pollution characteristics between 1987 and 1996. CZxSAX, north-west Bohemia, borders with Saxony; CZxPL, northern Moravia, Silesia.

The trend is complicated by a number of site-specific and annual phenomena. In 1996, although there was no increase in pollutant emissions, a moderate rise in pollutant concentrations, particularly of sulphur dioxide, was observed over most of the territory. It was caused by a long-term temperature inversion, although low SO_2 concentrations measured during the vegetative period indicate that the decrease in air pollution in the 1990s is long term.

Similar trends in sulphur dioxide emissions have been recorded in other CEE countries. Despite this, SO_2 is still a serious threat to the environment. In Poland, total annual emissions of SO_2 from all sources in the mid-1970s and throughout the 1980s ranged between 3 and 4.5 Mt year^{-1} (Fig. 8.3). In the 1990s, due to the lower industrial production and energy consumption, total emissions of SO_2 declined by almost 80% compared with 1978–1988 values. In the early 1990s it was estimated that in order to prevent further

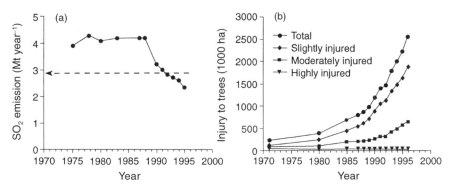

Fig. 8.3. (a) Estimated SO₂ emissions in Poland. (b) Area of injured forest in Poland as determined by defoliation (Ellsworth and Oleksyn, 1997, updated; Environment, 1997); the hatched line in (a) indicates the estimated requirements for the prevention of further deterioration of environmental conditions according to Kozlowski (1991).

deterioration of the environment it would be necessary to reduce SO₂ emissions to below 3 Mt by the year 2000 (Kozlowski, 1991). Current levels of SO₂ emissions are 22% lower than this target value. However, it is impossible to predict what future levels of emissions will be and how economic restructuring will affect pollution levels in the next decade.

Despite the decrease in SO₂ emissions throughout the region, there are striking differences in the emissions of post-Communist and developed countries, calculated on the basis of both per capita and per unit of gross national product (GNP; Fig. 8.4). SO₂ emissions calculated per GNP unit are 3 to 20 times higher in post-Communist countries than in the 15 European Union countries. The differences in the figures for NO$_x$ are much less pronounced. Recent improvements in the economies of some CEE countries have increased the buying power of citizens. This has resulted in a rapid increase in the number of private cars and, as a consequence, substantially increased emissions of ozone precursors (NO$_x$ and hydrocarbons). Although it is suspected that ozone concentrations have been steadily increasing in the region due to local formation and long-range transport from Western Europe, no thorough monitoring programme has been developed (Bytnerowicz, 1997). It is only since the 1990s that some pilot collaborative studies have been initiated in CEE countries.

There are two main patterns of ozone concentrations in Europe (Grennfelt and Beck, 1994): (i) latitudinal, with more frequent episodes of high ozone concentrations in Central Europe (south Germany, Switzerland, northern Italy) than in areas further to the north and west; and (ii) altitudinal, with increased incidents of high concentrations at higher elevations. The current threshold level for growth reductions in forest trees is an AOT40 (accumulated

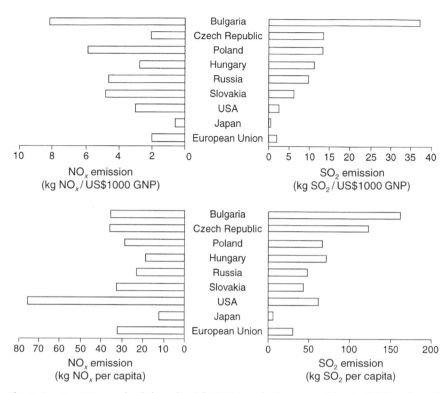

Fig. 8.4. Emissions of sulphur dioxide (SO_2) and nitrogen oxides (NO_x) in selected countries in 1994 or 1995 (Environment, 1997). GNP, gross national product.

exposure over a threshold of 40 ppb) value of 10 ppbh, calculated over a 6-month growing season (Fuhrer and Achermann, 1994). Current values exceed this threshold over most of the CEE countries (Hettelingh *et al.*, 1997).

Effects of pollution on forests

Elevated levels of toxic gases such as SO_2, NO_x, O_3 or HF resulting from industrial emissions have well-documented direct physiological effects on trees (Keller, 1984; Tjoelker *et al.*, 1995; Slovik *et al.*, 1996). After entering leaves through stomata these gases affect the mesophyll tissue, resulting in a reduction in the CO_2 assimilation and leaf life-span. Trees exposed to pollutant concentrations similar to those experienced by forests in industrialized areas can have growth reduced by up to 25% (Matyssek *et al.*, 1995).

Air pollution stress affecting forest trees also includes soil-mediated effects. Over the long term, the deposition of sulphur and nitrogen compounds can cause changes in soil properties. Three hypotheses concerning causal mechanisms dominate the relevant literature (Solberg and Torseth, 1997):

1. The load of strong acids may lead to losses of base cations (Ca and Mg) from the humus layer, causing development of nutrient deficiency. (Foliage yellowing and growth reductions due to Mg-deficiency have been induced in experiments with artificial acidification.)
2. Increased nitrogen input may contribute to nutrient deficiency and drought problems because it increases tree demands for other nutrients and water.
3. The acids may lead to lower pH values and thus increase availability of toxic aluminium in the soil solution.

Adverse effects of high Al concentration on trees have been demonstrated in both controlled and field experiments (Ulrich *et al.*, 1980; Rengel, 1992; Oleksyn *et al.*, 1996). For example in a heavily polluted region in Poland, high Al and low Mg concentrations in Scots pine (*Pinus sylvestris* L.) foliage was associated with reductions in photosynthesis of almost 40% and increases in needle respiration of more than 20% (Reich *et al.*, 1994). Stachurski *et al.* (1995), working with Norway spruce (*Picea abies* (L.) Karst.) in the Sudety Mountains (Poland), found that a major factor contributing to tree decline there is foliage deficiency of N, K, Ca, Mg and Fe, and imbalanced ratios of N : P, K : P, and F : P, together with toxic levels of Al, Pb and S. These factors reduce chlorophyll concentrations and promote the production of polyphenols. Magnesium deficiency is also a frequent cause of the decline of Norway spruce monocultures and stands growing on poor or degraded sites. The reduced vitality of forest stands and significant changes in the base element contents of the soil indicate a need for remedial measures. Liming using dolomite limestone has been used in many mountain areas since 1970 (Kubelka *et al.*, 1993), but there are both advantages and disadvantages associated with liming (see Kozlov *et al.*, this volume). Imbalanced nutrition should be considered as an important factor affecting tree health in CEE. It also can reduce resistance of forest ecosystems to combinations of other stresses.

The acidity of precipitation has decreased in the Czech Republic since 1993. However, as the buffering capacity of nutrient-poor forest soils has been lowered by past acidic deposition, leaching of nutrients continues. In Slovakia, total sulphur deposition still exceeds the critical loads for many ecosystems, ranging from 22 to 24 kg ha^{-1} year^{-1} (Caboun, 1999), although the critical level for sulphur dioxide is only exceeded locally. A similar situation exists with nitrogen oxides, but the monitoring network for NO_x is insufficiently dense to provide reliable results. Experimental measurements of ammonium have not indicated that critical levels are currently being exceeded. In future, hydrocarbons may become an important precursor for the formation of photochemical oxidants.

Critical levels of ozone are often exceeded during periods of unfavourable photochemical conditions. Preliminary results from the CEE countries suggest that high concentrations occur mainly in medium and high mountain regions, where O_3 concentrations exceeding the threshold level occur as early in the growing season as the end of May. For example, recent measurements conducted in southern Poland (Upper Silesia and Krakow regions, Beskid Mountains (known as the Beskidy Mountains in Poland and the Beskydy Mountains in the Czech Republic)) have shown that, 1-h average ozone concentrations in the summer may reach more than 120 ppb, leading to the occurrence of acute symptoms of ozone injury on the foliage of some woody plants (Bytnerowicz et al., 1993; Godzik, 1997). Ozone concentrations have not reached such high values in the Czech Republic, probably because of the wet and relatively cool weather in the summer months of recent years (Lomsky and Sramek, in press). However, taking into account possible future climatic changes and emissions scenarios, an average annual increase in O_3 concentrations of 0.5% can be expected (Caboun, 1999).

Both active and passive measurements of ozone, combined with recent monitoring using O_3-sensitive plants in some CEE countries, have confirmed that ozone should be considered as an important air pollutant affecting the vegetation of the region (Blum et al., 1997; Godzik, 1997; Zeller et al., 1997). However, the existing data are insufficient to draw any conclusions about the spatial distribution of ozone or the potential danger to any given country or region in the CEE countries. The ecological significance of long-term ozone exposures has been well documented in a variety of field studies and laboratory experiments (for reviews see Fuhrer and Achermann, 1994; Sandermann et al., 1997). It adversely influences stomatal conductance, photosynthesis and over the long term is believed to lead to growth reductions, water deficiency, reduced nutrient availability and a decrease in frost tolerance (Matyssek et al., 1991; Küppers et al., 1994; Tjoelker et al., 1995).

Combinations of air pollution and adverse meteorological conditions seem to be especially harmful to mountain forests throughout the region (Grodzinska and Szarek-Lukaszewska, 1997; Schwarz, 1997; Brázdil, 1998). For example, during the winter of 1995–1996, acute injury occurred to large areas of forests in the north-eastern parts of the Ore Mountains ridge, caused by long-term extreme conditions, high SO_2 concentrations and unfavourable climatic conditions (Fig. 8.5). The exceptional situation arose from the combination of high concentrations of phytotoxic substances and the presence of persistent rime. The damage was characterized by an extreme reddening of needles and the breakage of tree crowns and main branches. The damage mechanisms involved frost, extreme water stress as a result of foliar transpiration and the high volumes of ice in the crowns. The lowest temperatures were not particularly extreme (−17.4°C), but the freezing weather was prolonged. Unusually, the winds were continuously from the SE between November 1995 and February 1996, resulting in the large-scale transport of SO_2 into the

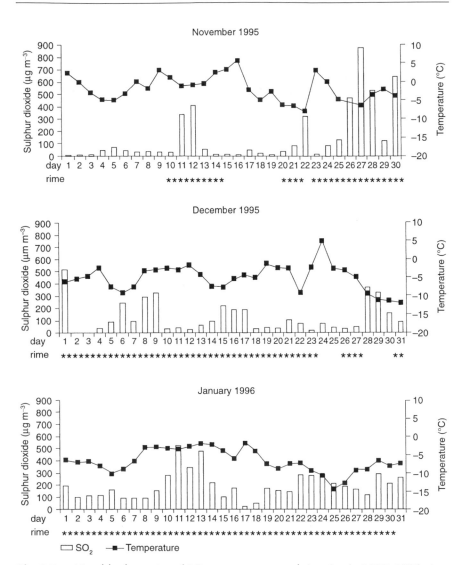

Fig. 8.5. Monthly dynamics of SO_2, temperature and rime ice in 1995–1996 at Medenec in the Ore Mountains, Czech Republic.

mountains, with peak concentrations in November 1995 (average daily concentrations of 1000 µg m⁻³). Peaks of SO_2 at Medenec between November and February were as follows: 1750 µg m⁻³ (27 November), 1538 µg m⁻³ (30 November), 817 µg m⁻³ (9 December), 872 µg m⁻³ (15 December), 2542 µg m⁻³ (28 December), 972 µg m⁻³ (11 January), 1289 µg m⁻³ (13 January), 906 µg m⁻³ (2 February) and 1396 µg m⁻³ (27 February).

The complex nature of forest decline is well illustrated by studies conducted in Saxony where forests are affected by sulphur, heavy metals and base cations from industrial sources and nitrogen from traffic and agricultural activities. During industrialization, SO_2 caused severe damage in the vicinity of point-sources. Since the end of the 1970s, a new type of forest decline has been detected. Defoliation of conifers has occurred at a regional scale (previously, needle loss had only been observed locally) and in combination with chlorosis of older needles, indicated magnesium deficiency as being involved. The causes of the deficiency were the high loads of acidifying and eutrophifying substances (Ulrich, 1993). Despite reductions in industrial emissions between 1989 and 1992, the acidity of precipitation continued to increase, probably because emissions in alkaline dusts were reduced faster than emissions of sulphur dioxide (Raben *et al.*, 1998). The fluxes of SO_4 in precipitation were lowered by 50% but concentrations of NO_x increased by about 30%.

The situation in Saxony is being investigated on six permanent survey plots, using the protocols of the International Cooperative Programme on the Assessment and Monitoring of Air Pollution Effects on Forests (ICP Forests). Values of the pH in throughfall dropped to 2.7 during the winter of 1995–1996 (Fig. 8.6, results from three plots along the Saxon–Czech border), in accordance with the high concentrations of SO_2 and NO_3 at that time (Raben *et al.*, 1996, 1998; LAF, 1997). In the 1996 hydrological year (11/95–10/96), sulphur deposition in throughfall at Olbernhau (Central Ore Mountains) amounted to 77 kg ha^{-1}, while the Norway spruce stand at Klingenthal (Western Ore Mountains) received 38 kg ha^{-1} of sulphur and 23 kg ha^{-1} of total nitrogen. Nitrogen was nearly equally divided into NH_4-N and NO_3-N. Inputs of free acidity (H^+) in the throughfall in the same year were about 4.0 kg ha^{-1} in the Norway spruce stand at Olbernhau.

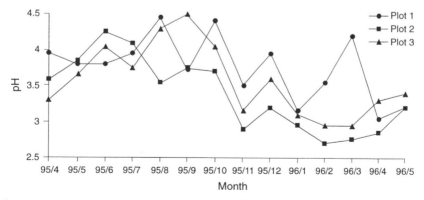

Fig. 8.6.   Throughfall pH at three monitoring plots in Saxony, April 1995 to May 1996. 1, Klingenthal (Erzgebirge W); 2, Olbernhau (M); 3, Cunnersdorf (E).

Stress indicators in the soil solution (molar ratios of Ca/H, Ca/Al, Mg/Al) indicate further acidification of soil water. In combination with periods of strong frost, severe damage has occurred to Norway spruce stands over about 50,000 ha and the dieback of 3000 ha has been documented in the Saxon area of the Central Ore Mountains during the winter and spring of 1996. At the same time, damage to about 30,000 ha has been reported from the Czech side of the Ore Mountains ridge. The damage is considered to be serious in 16,000 ha, and has resulted in the loss of 2800 ha of Norway spruce, mainly in young stands.

In the past, trees believed to be more tolerant of environmental pollution ('substitute species') were widely used in plantations of 'transitory stands' under the assumption that they would ensure the environmental and to a certain extent also the production functions of forest ecosystems in heavily polluted regions. For example, in the Ore Mountains, with the exception of introduced spruces such as Colorado blue spruce (*Picea pungens* Parry ex Engelm.) and black spruce (*Picea mariana* (Mill.) BSP), silver birch (*Betula pendula* Roth.) is the major substitute species over vast areas. However, in 1997, out of a total area of 12,000 ha planted with birch, nearly 3000 ha were damaged. An initial survey has excluded SO_2 and NO_x as possible causes of the damage and has suggested that other pollutants present during bud development in 1996, and physiological drought, predisposed the trees to decline (Lomsky and Sramek, in press).

Despite the many problems described above, growth reductions have only been noted near point-sources of pollution or at the regional scale. On average, growth rates of trees and stands in Europe are currently higher than have been recorded at any time in the past (Kauppi *et al.*, 1992; Kandler and Innes, 1995; Spiecker *et al.*, 1996; Ellsworth and Oleksyn, 1997). The reasons for this are uncertain, although increases in forest area have not substantially contributed to the observed trends. It is possible that high atmospheric nitrogen deposition and increasing atmospheric CO_2 concentrations are partially responsible for the growth enhancement, and that these factors may have mitigated the effects of other environmental pollutants by stimulating tree growth (Kauppi *et al.*, 1992; Ellsworth and Oleksyn, 1997).

Forest monitoring is undertaken in all the countries of the region. Defoliation is assessed annually on many thousands of trees, using national forest health surveys and those conducted under the guidelines of the International Cooperative Programme for the Assessment and Monitoring of Air Pollution Effects on Forests (Anon., 1998). Defoliation levels have not changed consistently in the different countries of the region over the decade measurements have been undertaken, although a general deterioration in crown condition has been detected. Crown condition scores are affected by the methodology adopted in individual countries and by local conditions. For example, stand age plays an important role in the Czech Republic, where it is mostly older trees that are observed. This may be one of the reasons that the highest scores (70%

of trees) for the 'moderate defoliation' category (25-60%) were recorded in 1996. In marked contrast, the scores for this category in 1996 were 37% in Poland and 30% in Slovakia (Anon., 1997) (Table 8.1).

An improvement in forest condition (as determined by defoliation) has been noted in the eastern part of Germany in the first half of the 1990s and also in some eastern European countries. It is likely that slightly higher precipitation levels during the vegetation periods in 1995 and 1996 benefited the health of forest stands as the period 1985–1994 was marked by hot and dry weather during the spring and summer. However, despite marked reductions in pollutant emissions, particularly sulphur dioxide, air pollution continues to impact large areas of Eastern Europe (Raben et al., 1996).

There is no correlation between air pollution concentrations and tree health, as defined by defoliation, at the national level, e.g. in Slovakia (Caboun, in press) or Poland, or at the regional level (Kandler and Innes, 1995; Bussotti and Ferretti, 1998). Defoliation data alone are unlikely to provide sufficient information about the effects of air pollution on the condition of forest stands. Consequently, to determine cause–effect relationships, more information is required. This includes data on atmospheric deposition, soil chemistry, air quality and climate, together with information on the many factors affecting the condition of particular forest stands in the region (including past mismanagement, browsing intensity, etc.). For example, salvage fellings can represent a substantial part of the annual cut in some countries (in the Czech Republic, salvage cutting accounted for 77.7% of the harvest in 1994, 63.5% in 1995 and 60% in 1996) (Fig. 8.7) and the application of effective forest management in such cases can be difficult (Vasicek, 1997).

Along with pollution, plants are usually exposed to combinations of other stress factors. The importance of biotic and abiotic interactions in controlling plant response and variability in space and time are the two most important

Table 8.1. Defoliation classes 2–4(%); according to ICP Forests (+ worsening).

Country/year	1993	1994	1995	1996	1997	1993–97 +	−
Czech Republic	51.8	57.7	58.5	71.9	68.6	16.8	
Germany	24.2	24.4	22.1	20.3	19.8		4.4
Poland	50.0	54.9	52.6	39.7	36.6		13.4
Slovakia	37.6	41.8	42.6	34.0	31.0		6.6
Belarus	29.3	37.4	38.3	39.7	36.3	7.0	
Ukraine	21.5	32.4	29.6	46.0	31.4	9.9	
Hungary	21.0	21.7	20.0	19.2	19.4		1.6
Slovenia	19.0	16.0	24.7	19.0	25.7	6.7	
Croatia	19.2	28.8	39.8	30.1	33.1	13.9	
Romania	20.5	21.2	21.2	16.9	15.6		4.9
Bulgaria	23.2	28.9	38.0	39.2	49.6	26.4	

Fig. 8.7. Damaged Norway spruce forest in the Giant Mountains, northern Czech Republic.

reasons for the lack of success in improving the understanding of the long-term consequences of air pollution stress on forest productivity, community dynamics and biogeochemistry (Taylor *et al.*, 1994). Unfortunately, the literature on the interacting effects of different types of stresses and pollutants on trees is scanty. Below, we describe the two most important groups of factors which alone or in combination with pollution threaten CEE forests.

2.2   Climatic threats

The nature and occurrence of the abiotic factors which threaten forests are largely determined by climate. The climatic characteristics of Central Europe result in weather anomalies and in recent years there have been more frequent extreme temperatures, precipitation and winds (Brázdil, 1998). In contrast to the physical separation of the Czech Republic from Slovakia by a series of mountain ridges, there are no orographic barriers to air masses penetrating from the western Atlantic. These air masses, which reach the eastern part of Germany, Poland, the Ukraine and Belarus, transport pollutants eastwards from the most industrialized regions of Western Europe.

Wind, snow and frosts are the main abiotic stresses affecting forests. Storms have often caused disastrous damage, and their occurrence in the region appears to be becoming more and more frequent, such that the total volume of damaged timber has reached very high figures. For example, the highest proportion of salvage felling in Slovakia is caused by wind-blown stands (67%) and only 13% arises from damage caused by air pollution. Substantial damage by snow has also occurred several times in recent decades (Vasicek, 1997).

Frost can have serious consequences in the region, such as on the night of 31 December 1978 – 1 January 1979, when the temperature dropped during a period of 6–8 hours from +8°C to −23°C. There have also been problems associated with frost damage to newly flushed foliage, as in the beech (*Fagus sylvatica* L.) forests of the Beskid Mountains (Vancura, 1992). A special case was the winter of 1995–1996, when low temperatures alternating with strong frosts combined with long-term exposure to rime, fog and air pollution to cause severe damage, particularly in the Ore (Erzgebirge) Mountains. The prevailing cold and wet weather in the spring and summer of 1995 and 1996 adversely influenced the reproduction of a number of insect pests, but the wet weather favoured the spread and increased frequency of a number of pathogenic fungi (Zahradnik, 1997). Drought has also been a cause of tree mortality, and not only on sandy sites. The last decade has been very dry but precipitation levels in 1996 and 1997 were close to normal. Droughts can have direct effects, but can also contribute to the activation of other disease processes (Gorzelak, 1995).

2.3   Biotic threats

Changes in the occurrence of biotic threats result mainly from variations in weather conditions and the consequences of human activities, including air pollution and unsuitable forest management practices. Trees are often attacked by secondary insect pests and parasitic fungi following abiotic and anthropogenic stresses. Forest diseases in the region are associated with the high frequency of pest species characterized by large-scale population fluctuations and with the occurrence of parasitic fungi. For example, infections by the pine needle cast *Lophodermium* spp. P. Karst have been unusually intensive in recent years. Other problems include the decline of young larch stands, and declining oak stands, although these seem to have stabilized. In parts of the region (Czech Republic, Germany), oaks (*Quercus* spp.) have been severely attacked by oak leaf roller moth (*Tortrix viridana* L.) and winter moth *Operophtera* spp. (Zahradnik, 1997).

At the start of the 1990s, the most significant defoliating insects on coniferous trees were three pests of Norway spruce: the nun-moth (*Lymantria monacha* L.), the sawfly (*Cephalcia abietis* L.) and the gregarious spruce sawfly (*Pristiphora abietina* Christ), all of which have been suggested as possible bioindicators of air pollution (Gorzelak, 1995). The nun-moth represents a serious problem mainly in Poland, Bohemia and Moravia; its most recent outbreak affected a peak of 20,000 ha, but declined rapidly in 1996 to 1700 ha of affected forest. In the Czech Republic the greatest amount of damage is still being caused by bark beetles. In 1996, the total volume of spruce wood infested by bark beetles was 967,000 m^3, roughly half the total in 1995. The most serious is the larger eight-toothed spruce bark-beetle (*Ips typographus* L.), although the pine bark-beetle (*Pityogenes chalcographus* L.), small spruce bark-beetle (*Polygraphus poligraphus* L.) and Nordic bark-beetle (*Ips duplicatus* Sahlb.) are also important. The Nordic bark-beetle has become a problem as a result of increasing international timber trade and inappropriate management practices in the newly privatized forests. In northern Moravia and Silesia, new species probably entered the Czech Republic via timber imports from Belarus and Poland. Generally, the greatest problems are experienced on private forest properties, particularly smaller ones (Zahradnik, 1997).

Outbreaks by rodents (common vole, *Microtus arvalis* L. and bank vole *Cletrionomys glareolus* L.) and the damage that they cause have generally been low in recent years. This may be due to natural factors and/or decreasing reforestation rates in mountain areas (Zahradnik, 1997). Ungulates, including red deer (*Cervus elaphus* L.), mouflon (*Ovis musimon* L.) and others, are still among the most significant factors damaging forests in the Czech Republic, Slovakia and also partly in Poland. This damage is considered to be anthropogenic in the Czech Republic because of the artificially high populations, despite recent decreases in the abundance of red deer (Zahradnik, 1997).

3   Case examples of regional pollution effects on forests

3.1   Sudety Mountains forest decline ('Black Triangle' region, south-west Poland)

The forests of the Sudety Mountains, Poland, serve as an example of historical and recent anthropogenic changes in forest condition. The area of forest was reduced by half between the 12th and 18th centuries. Oak (*Quercus* spp.), hornbeam (*Carpinus betulus* L.), beech (*Fagus sylvatica* L.) and silver fir (*Abies alba* Miller) were among the most heavily harvested tree species. Attempts to begin forest management in this region started in the early 1810s, when forest cuts were limited in size and the seed trees had to be left. However, due to the high demand for spruce wood, all hardwood stands were systematically cut and replaced with Norway spruce. As a result, Norway spruce had become an important tree species in the Sudety Mountains by the end of the 19th century. Between 1834 and 1925 the proportion of Norway spruce increased from 30 to 95%, with the remainder made up of Scots pine (*Pinus sylvestris* L.) and European beech. Simultaneously, the standing biomass of the forest systematically declined from 527 m^3 ha^{-1} in 1775 to 353 m^3 ha^{-1} in 1890 and 215 m^3 ha^{-1} in the mid-1980s (Wiczkiewicz, 1982; Mazurski, 1986).

Natural regeneration of Norway spruce was successful in only a few places, and by the end of the 19th century local seed sources provided insufficient quantities of seeds. Consequently, additional seeds were purchased from Austrian dealers. At that time the significance of the seed source for productivity and the resistance of plants to biotic and abiotic factors had not yet been recognized. This period of forest management thus led to the formation of monospecific, even-aged Norway spruce plantations, sensitive to many biotic and abiotic factors.

Pollution impacts on the Norway spruce forests of the Sudety Mountains were noted early in the 20th century in areas surrounding the coal mines and other small industrial plants in mountain valleys (Capecki, 1989). After World War II, rich deposits of soft (brown) coal were discovered in the region where the borders of south-west Poland, Germany and the Czech Republic meet. This region represents the largest basin of soft coal in Europe, with approximately 200–220 Mt being extracted annually (25% of the total production in Europe). The coal is burned locally in 17 major power plants with a total output over 15,000 MW. As a result, in this relatively small area (less than a quarter of the territory of the Netherlands), 3 Mt of SO_2 and approximately 1 Mt of NO_x were emitted each year. Adverse effects of air pollution developed throughout the Sudety region (in all three countries). Large-scale effects of pollution had already been detected at the end of the 1960s (Capecki, 1969), and in the 1980s large-scale forest disturbances (mostly involving Norway spruce) affected more than 46,000 ha in Poland. In 1984 alone in just one small part of the Sudety Mountains, forest decline affected 3800 ha of Norway spruce. It was the first such large decline in Polish mountain forests.

It is generally believed that a combination of air pollution, the larch bud moth (*Zeiraphera griseana*), and a secondary insect pest, *Ips typographus*, were responsible for the forest decline. Initially, only forests above 900 m elevation were affected but, over time, the problem occurred as low as 700 m elevation. During the peak period of tree mortality, over 1000 workers were required daily to remove dead wood in the affected Forest Districts in Poland (Oleksyn and Reich, 1994). From the mid-1980s until the present, forest dieback and mortality have continued in the Polish Sudety range, and shown up in increasingly young stands, without being accompanied by further insect outbreaks. This trend suggests that pollution may have been the primary cause of the earlier dieback and decline, which was then further exacerbated by the larch bud moth and secondary agents.

Reforestation of affected sites was attempted with the tree species used in the reforestation of lowland polluted sites in Poland, but these plantings failed (Oleksyn and Reich, 1994). A dense thatch cover established after the trees died and now impedes planting. High mice populations are also a problem. Even with unlimited funds, successful restoration of forest ecosystems in this degraded landscape will probably require further reductions in pollution levels and the likely 'sacrifice' of the first tree generation as a cover crop. The eventual reconstruction of these devastated forest stands will require a change of stand composition from Norway spruce to more natural mixed stands with hardwood species, including European beech. The extreme meteorological conditions that exist in high elevation forests, stand history (planting of exotic genotypes) and pollution probably combined to make these forests more susceptible than forests in other, more polluted, areas.

3.2 Effects of radioactive contamination on forests following the Chernobyl disaster

As a result of the explosion and subsequent fires in the reactor of the Chernobyl, Ukraine, nuclear power plant between 26 April and 5 May 1986, between 3 and 10% of the reactor's total radioactivity or about 25–50 million curies (1 Ci = 3.7×10^{10} Bq) of radioactive elements were released into the environment (Savchenko, 1995). The isotopes from the fallout consist of about 30 radionuclides with a half-life ranging from *c.* 3 days (Np-139, Mo-99, Te-132) to 24,110 years for plutonium (Pu-239). The main harm to the local ecosystems is from the long-lived radionuclides caesium-137 (half-life 30 years), strontium-90 (28 years) and plutonium nuclear fuel 'hot particles' (Myttenaere *et al.*, 1993).

After the catastrophe, large areas of Belarus, the Russian Federation and the Ukraine were contaminated. It is estimated that over 28,000 km^2 are contaminated with more than 5 Ci km^{-2}, and 76,100 km^{-2} with 1–5 Ci km^{-2}. In the three countries, a total of over 3.5 Mha of forests have been affected (Pisarenko *et al.*, 1994). Radiation control units of the forest administration in the Russian Federation used ground surveys to identify 958,700 ha of forests with a caesium-137 density exceeding 1 Ci km^{-2} (Shubin, 1996). In addition, the State Committee of Hydrology and Meteorology identified radioactive contamination in Tatarstan, Chuvashia and Saratov, and Nizhny Novgorod provinces. Very considerable differences in the pollution level are observed within relatively small areas, and this has necessitated the mapping of the pollution in forests at the sub-compartment level (Kaletnik, 1990).

Studies conducted in affected forests have revealed that any changes in radioactivity can be attributed almost exclusively to decay, and that it is almost impossible to stimulate radionuclide removal by applying any technical means. Data collected after the Chernobyl and Kyshtym (Russian Federation) accidents in the Ural Mountains suggest that forest ecosystems are effective in limiting the further spread of contamination away from the point of initial deposition and that this effect will increase over time (Tikhomirov *et al.*, 1993a,b). Different ecosystems absorb radionuclides in different ways, and forest ecosystems have been found to be more contaminated than other ecosystems (Savchenko, 1995). This is mainly due to significant absorption from the atmosphere of radioactive carbon-14 and tritium during photosynthesis.

The radionuclides then reach the soil via litterfall, where they accumulate in the litter and humus layers (Poiarkov *et al.*, 1995). Earthworm activity results in some of the radionuclides being translocated down the soil profiles. The grubbing and rooting activity of wild boar (*Sus scrofa* L.) is also a significant factor affecting the redistribution and migration of radioactive substances in forests (Lysikov, 1997).

There are mixed reports regarding the migration rates of radioactive elements in different forest types. Shcheglov *et al.* (1992) found that in purely coniferous stands the migration rate was lower than in deciduous forest stands or mixed stands. However, Tikhomirov *et al.* (1991) consider that the migration of all radionuclides, especially Ru-106, is greater in coniferous forest soils than in deciduous and mixed forest soils. Due to the longer foliage lifespan in evergreen coniferous species, radionuclides are retained longer in the canopy than in deciduous forests. Depending on the forest type involved, the litter contains 45–70% of the radioactive fallout, the tree layer 2–12% and the mineral soil 20–40% (Tikhomirov *et al.*, 1994).

In the zone of highest radiation (within a 30 km radius of Chernobyl) all forestry activity has been stopped, and 15 million m^3 of growing timber has been withdrawn from economic use (Kaletnik, 1990). Various anatomical and morphological changes in the foliage, shoots, seed germination energy and germinating ability were observed in the first growing season following the disaster (Abaturov *et al.*, 1992; Yushkov *et al.*, 1994). The needles of conifers suffered severe mortality (Izrael *et al.*, 1988) and later, numerous mutations were recorded in Norway spruce (Ipatjew *et al.*, 1994). The tree crowns in the high-impact zone initially retained between 60 and 90% of the radioactive fallout (Mamikhin *et al.*, 1994). Subsequently, self-cleansing of the crowns reduced this load to *c.* 18% by August 1986 and to 10% by July 1987. Very different patterns occurred with the contamination of wood by Cs-137. Higher levels of radiation were found at stand margins and in gaps, roadways and thinned areas (Lysikov, 1992).

The removal of surface litter falling 1–3 years after the disaster was recommended as an effective pollution control measure, in contrast to burial in deep trenches, which resulted in Sr-90 and Cs-137 quickly reaching the groundwater (Tikhomirov, 1993). It was also proposed that former agricultural land within the Chernobyl exclusion zone should be afforested with conifers, which at maturity (80–100 years) will provide wood which can be used without radiation hazard. Scots pine plantations have been established on about 150,000 ha of land that has been withdrawn in Belarus from agricultural cropping because of radioactive contamination (Rakhteenko, 1995). Pine growth is only significantly affected in areas with very high contamination (380–550 Ci km^{-2}), and this level has been taken as a threshold above which it is undesirable to plant Scots pine.

In almost all contaminated areas, the radioactivity levels in forest berries and mushrooms, which are an important component of the diet of local inhabitants, exceed permissible standards. The radioactivity of birch sap and medicinal plants also exceed the standards, although the contamination of the produce is dependent on the level of background contamination of the forest land (Los' *et al.*, 1991). Different species of edible mushrooms accumulate radionuclides at differing rates. In areas with contamination below 5 Ci km^{-2}, collection of species with low accumulations of Cs-137 (e.g. *Boletus scaber*, *Boletus edulis*, *Armillaria mellea*, *Gyromitra esculenta*, *Cantharellus cibarius*, *Psalliotia sylvatica*) is permitted after checking their radioactivity (Shubin, 1995). No collecting is allowed of species belonging to the groups of inter-mediate- or high-accumulators or in areas with a Cs-137 density exceeding 5 Ci km^{-2}. However, it is impossible to implement these restrictions fully. Dietary surveys of more than 700 food samples and whole human body moni-toring for Cs-137 conducted in the Bryansk Region of the Russian Federation in 1994–1995 found that radioactivity of wild food products, especially mushrooms, was higher than in agricultural products (Skuterud *et al.*, 1997). Consumption of mushrooms was also the main reason for a 60–70% increase in radioactivity concentrations in humans in autumn.

In areas with a radiation density above 15 Ci km^{-2} (more than 10,000 km^2 in Belarus, Russian Federation and the Ukraine), where the Cs-137 concentration in bark and logging residues exceeds permitted limits, logging is restricted to winter when the ground is snow-covered and opera-tions therefore do not disturb the soil and litter. Bark and slash must be left in the forest. The use of any forest products (except wood without bark) is forbidden. In zones with a radiation density above 40 Ci km^{-2} (more than 3000 km^2), wood radiation is more than 10,000 Bq kg^{-1} and no forestry activity is allowed, apart from fire-control measures (Pisarenko *et al.*, 1994; Shubin, 1995). As with mushrooms, tree species differ in their accumulation rates of radionuclides. In the Bryansk region in 1988–1990, the content of Cs-137 in wood decreased in the following order: aspen (*Populus tremula*), small-leaved lime (*Tilia cordata*), alder (*Alnus* spp.), oak (*Quercus robur*), birch (*Betula* spp.), Norway spruce and Scots pine (Ushakov *et al.*, 1992).

An especially dangerous situation can be created by forest fires in contam-inated regions. Radioactive smoke can be carried up to 25 km from the fire and create secondary contamination. In addition, the radioactivity of the ash (from 300 to 500 kg ha^{-1}) is equal or close to that of industrial radioactive waste.

Radioactive contamination of the forest also creates social problems and health hazards. For example in the Bryansk region (south-west Russia), loggers have radiation doses 3–5 times higher than those of other categories of workers (Boginskii *et al.*, 1996). There is also a tremendous financial burden imposed on affected countries resulting from:

- the withdrawal from economic use of wood and secondary products (berries, mushrooms, birch sap, medicinal plants);
- the need for the implementation of strict and effective fire-control measures;

- the establishment of a network of radiation monitoring stations and special laboratories to study the effects of the radioactivity on forests and the migration of the radionuclides;
- the extra costs involved in carrying out practical forestry operations in the various radiation zones (from 1.4 to 10.7 times the normal level, depending on the zone) (Kaletnik, 1990; Myttenaere et al., 1993; Abduragimov and Odnol'ko, 1994; Maradulin et al., 1996).

4 Conclusions

Due to the combination of economic restructuring, production declines and technological changes since the mid-1980s in the countries of CEE, a considerable decline in emissions and the deposition of airborne pollutants has occurred. However, contrary to most expectations, defoliation levels have not changed substantially. The apparent lack of a large-scale relation between pollution load, forest productivity and crown condition should be further investigated.

Despite the significant decreases in the current levels of emissions, acidifying substances such as SO_2 and NO_x remain a problem. Emissions need to be further reduced, both locally and internationally, and the nutrient status of forest soils needs to be improved. Combinations of extreme climatic conditions, biotic factors and air pollution still affect substantial areas of CEE where local and regional forest decline is observed. Due to significant changes in the pollution climate (decreased SO_2 and increased O_3) and possible climatic change there is an urgent need to establish a reliable forest monitoring network throughout the entire region. It is also necessary to adopt and implement sound management practices as well as to create financial incentives for environmental protection.

Large forested areas of Belarus, Ukraine and the Russian Federation are contaminated with radioactive substances after the Chernobyl and other nuclear accidents. The contamination is creating a variety of management and economic challenges for these countries. It also provides a unique opportunity to address numerous scientific problems and to learn how to cope with future nuclear contamination of forests.

References

Abaturov, Yu. D., Gol'tsova, N.I., Rostova, N.S., Girbasova, A.V., Abaturov, A.V. and Melankholin, P.N. (1992) Some features of radiation damage to pine in the region of the Chernobyl disaster. *Soviet Journal of Ecology* 22, 298–302.
Abduragimov, I.M. and Odnol'ko, A.A. (1994) The Chernobyl echo of forest fires. *Lesnoe Khoz* 2 (1994), 30–32 (in Russian).

Anonymous (1997) Forest condition in Europe. Results of the 1997 crown condition survey. UN Economic Commission for Europe, Geneva, and European Commission, Brussels.

Anonymous (1998) *Manual on Methods and Criteria for Harmonized Sampling, Assessment, Monitoring and Analysis of the Effects of Air Pollution on Forests*, 4th edn. Federal Research Centre for Forestry and Forest Products, Hamburg, in 8 parts.

Bialobok, S. (ed.) (1989) *Tree Life in a Polluted Environment*. PWN, Warsaw-Poznan (in Polish with English summaries).

Blum, O., Bytnerowicz, A., Manning, W. and Popovicheva, L. (1997) Ambient tropospheric ozone in the Ukrainian Carpathian Mountains and Kiev region: detection with passive samplers and bioindicator plants. *Environmental Pollution* 98, 299–304.

Boginskii, N.I., Aleshchenko, V.M., Mishin, S.V. and Panaskin, V.V. (1996) Prophylactic measures against fires and radiological monitoring in the forests of the Bryansk region. *Lesnoe Khoz* 6 (1996), 25–26 (in Russian).

Brázdil, R. (1998) Meteorological extremes and their impacts on forests in the Czech Republic. In: Beniston, M. and Innes, J.L. (eds) *The Impacts of Climate Variability on Forests*. Springer-Verlag, Berlin, pp. 19–47.

Bussotti, F. and Ferretti, M. (1998) Air pollution, forest condition and forest decline in Southern Europe: an overview. *Environmental Pollution* 101, 49–65.

Bytnerowicz, A. (1997) Air pollution and climate change effects on forests in Central and Eastern Europe. *Environmental Pollution* 98, 271.

Bytnerowicz, A., Manning, W.J., Grosjean, D., Chmielewski, W., Dmuchowski, W., Grodzinska, K. and Godzik, B. (1993) Detecting ozone and demonstrating its phytotoxicity in forested areas of Poland: a pilot study. *Environmental Pollution* 80, 301–305.

Caboun, V. (1999) *Impacts of Air Pollution in Slovakia*. VULHM, Zpravy.

Capecki, Z. (1969) Danger to the Sudety forests by pests on the background of forest decline cost by hurricanes and mires. *Sylwan* 3, 57–64 (in Polish).

Capecki, Z. (1989) Forest condition of different parts of the Sudety forests. *Prace IBL* 688, 5–93 (in Polish).

Ellsworth, D.S. and Oleksyn, J. (1997) Evaluating of air pollution to forests in Central and Eastern Europe. In: Gutkowski, R.M. and Winnicki, T. (eds) *Restoration of Forests. Environmental Challenges in Central and Eastern Europe*. Kluwer Academic, Dordrecht, pp. 121–131.

Environment (1997) *Information and Statistical Papers*. Central Statistical Office, Warsaw (in Polish).

Forestry (1997) *Information and Statistical Papers*. Central Statistical Office, Warsaw (in Polish).

Fuhrer, J. and Achermann, B. (eds) (1994) *Critical Levels for Ozone. A UN-ECE Workshop Report*. Swiss Federal Research Station for Agricultural Chemistry and Environmental Hygiene, Liebefeld-Bern.

Godzik, B. (1997) Ground level ozone concentrations in the Krakow region, southern Poland. *Environmental Pollution* 98, 273–280.

Godzik, S. and Sienkiewicz, J. (1990) Air pollution and forest health in Central Europe: Poland, Czechoslovakia and the German Democratic Republic. In: Grodzinski, W., Cowling, W.B. and Breymayer, A.I. (eds) *Ecological Risks – Perspectives from*

Poland and the United States. National Academy Press, New Haven, Connecticut, pp. 22–74.

Gorzelak, A. (1995) Forests and Forest Management in the Sudety Mountains. *Works of the Forest Research Institute* Series B, 25/1, 7–35 (in Polish).

Grennfelt, P. and Beck, J.P. (1994) Ozone concentrations in Europe in relation to different concepts of the critical level. In: Fuhrer, J. and Achermann, B. (eds) *Critical Levels for Ozone. A UN-ECE Workshop Report.* Swiss Federal Research Station for Agricultural Chemistry and Environmental Hygiene, Liebefeld-Bern, pp. 184–194.

Grodzinska, K. and Szarek-Lukaszewska, G. (1997) Polish mountain forests: past present and future. *Environmental Pollution* 98, 369–374.

Hettelingh, J.-P., Posch, M. and de Smet, P.A.M. (1997) Analysis of European maps. In: Posch, M., Hettelingh, J.-P., de Smet, P.A.M. and Downing, R.J. (eds) *Calculation and Mapping of Critical Thresholds in Europe. Status Report 1997.* Coordination Centre for Effects, RIVM Report No. 259101007. National Institute of Public Health and the Environment, Bilthoven, pp. 3–18.

Innes, J.L. (1993) Air pollution and forests – an overview. In: Schlaepfer, R. (ed.) *Long-term Implications of Climate Change and Air Pollution on Forest Ecosystems,* IUFRO World Series, Vol. 4. IUFRO, Vienna, pp. 77–100.

Ipatjew, W.A. and Ignattschik, V.W. (1994) Einfluss des Tschernobyl-Reaktorunfalls auf die Wälder Weissrusslands. *Allgemeine Forstzeitschrift* 49, 662–666.

Izrael, Yu.A., Sokolovskii, V.G., Sololov, V.E., Vetrov, V.A., Dibobes, I.K., Trusov, A.G., Ryabov, I.N., Aleksakhin, R.M., Povalyaev, A.P., Buldakov, L.A. and Borzilov, V.A. (1988) Ecological consequences of radioactive contamination of natural environments in the region of Chernobyl accidents. *Atomnaya Energiya* 64, 28–40.

Kaletnik, N.N. (1990) Organization of forestry in conditions of radioactive pollution of the forests of the Ukrainian Poles'e. *Lesnoe Khoz* 4 (1990), 30–31.

Kandler, O. and Innes, J.L. (1995) Air pollution and forest decline in Central Europe. *Environmental Pollution* 90, 171–180.

Kauppi, P.E., Mielikäinen, K. and Kuusela, K. (1992) Biomass and carbon budget of European forests, 1971 to 1990. *Science* 256, 70–74.

Keller, T. (1984) Direct effects of sulfur dioxide on trees. *Philosophical Transactions of the Royal Society of London,* Series B 305, 317–329.

Kozlowski, S. (1991) *Industry and Environment.* PWN, Warsaw (in Polish).

Kubelka, L., Karásek, A., Rybář, V., Badalík, V. and Slodičák, M. (1993) *Forest Regeneration in the Heavily Polluted NE 'Krusne hory' Mountains.* Czech Ministry of Agriculture, Prague.

Küppers, K., Boomers, J., Hestermann, S., Hanstein, S. and Guderian, R. (1994) Reaction of forest trees to different exposure profiles of ozone dominated air pollution mixtures. In: Fuhrer, J. and Achermann, B. (eds) *Critical Levels for Ozone. A UN-ECE Workshop Report.* Swiss Federal Research Station for Agricultural Chemistry and Environmental Hygiene, Liebefeld-Bern, pp. 98–112.

LAF (Sächsische Landesanstalt für Forsten) (1997) *Waldschadensbericht 1997.* SMLEF, Graupa.

Lomsky, B. and Sramek, V. (1999) The Ore Mountains 1997: Experiences and New Problems. Proceedings of international workshop, Kovarsk· 9/97. *Zpravy Lesnickeho Vyzkumu* [*Forestry*].

Los', I.P., Tereshchenko, V.M., Perevoznikov, O.N., Kaletnik, N.N., Landin, V.P., Krasnov, V.P. and Maradudin, I.I. (1991) The radiation situation in the forestry enterprises of the Ukraine operating in conditions of radioactive contamination. *Lesnoe Khoz* 6 (1991), 28–29 (in Russian).

Lysikov, A.B. (1992) The radiation situation and distribution of radionuclides in Scots pine stands in the Chernobyl nuclear power station disaster zone. *Lesovedenie* 3 (1992), 52–60 (in Russian with English summary).

Lysikov, A.B. (1997) The radiation situation of contaminated pine stands in the south-western part of the Bryansk region. *Lesovedenie* 4 (1997), 71–79 (in Russian with English summary).

Mamikhin, S.V., Tikhomitov, F.A. and Shcheglov, A.I. (1994) Dynamics of ^{137}Cs content in forest biogeocenoses subjected to radioactive contamination as a result of the Chernobyl accident. *Russian Journal of Ecology* 25, 106–110.

Maradulin, I.I., Panfilov, A.V. and Rusina, T.V. (1996) Russian forestry as affected by radioactivity. *Khimiya v Sel'sk Khoz* 1 (1996), 11–13 (in Russian).

Matyssek, R., Gunthard-Goerg, M.S., Keller, T. and Scheidegger, C. (1991) Impairment of gas exchange and structure in birch leaves caused by low ozone concentrations. *Trees* 5, 5–13.

Matyssek, R., Reich, P.B., Oren, R. and Winner, W.E. (1995) Response mechanisms of conifers to air pollutants. In: Smith, W.K. and Hinckley, T.M. (eds) *Ecophysiology of Coniferous Forests*. Academic Press, San Diego, pp. 255–308.

Mazurski, K.R. (1986) The destruction of forests in the Polish Sudetes Mountains by industrial emissions. *Forest Ecology and Management* 17, 303–315.

Moldan, B. (1997) Czech Republic. In: Klarer, J. and Moldan, B. (eds) *The Environmental Challenge for Central European Economies in Transition*. John Wiley & Sons, Chichester, pp. 107–129.

Myttenaere, C., Schell, W.R., Thiry, Y., Sombre, L., Ronneau, C. and Schrieck, J. (1993) Modelling of Cs-137 cycling in forests: recent developments and research needed. *Science of the Total Environment* 136, 77–91.

Oleksyn, J. and Reich, P.B. (1994) Pollution, habitat destruction, and biodiversity in Poland. *Conservation Biology* 8, 943–960.

Oleksyn, J., Karolewski, P., Giertych, M.J., Werner, A., Tjoelker, M.G. and Reich, P.B. (1996) Altered root growth and plant chemistry of *Pinus sylvestris* seedlings subjected to aluminum in nutrient solution. *Trees* 10, 135–144.

Pisarenko, A.I., Sidorov, V.P., Tikhomirov, F.A., Chilimov, A.I. and Shcheglov, A.I. (1994) Principal positions for the practice of forestry in conditions of radioactive pollution. *Lesnoe Khoz* 2, 5–8 (in Russian).

Poiarkov, V.A., Nazarov, A.N. and Kaletnik, N.N. (1995) Post-Chernobyl radio-monitoring of Ukrainian forest ecosystems. *Journal of Environmental Radioactivity* 26, 259–271.

Raben, G., Andrae, H. and Leube, F. (1996) Schadstoffbelastungen in sächsischen Waldökosystemen. *Allgemeine Forstzeitschrift/Der Wald* 51, 1244–1248.

Raben, G., Andrae, H. and Symossek, F. (1998) Consequences of reduced emissions on the ecochemical conditions of forest ecosystems in Saxony (Germany). *Chemosphere* 36, 1007–1012.

Rakhteenko, L.I. (1995) Growth of pine plantations in the fallout zone of Chernobyl emissions. *Lesnoe Khoz* 1, 36–37 (in Russian).

Reich, P.B., Oleksyn, J. and Tjoelker, M.G. (1994) Relationship of aluminium and calcium to net CO_2 exchange among diverse Scots pine provenances under pollution stress in Poland. *Oecologia* 97, 82–92.

Rengel, Z. (1992) Role of calcium in aluminium toxicity. *New Phytologist* 121, 499–513.

Sandermann, H., Wellburn, A.R. and Heath, R.L. (eds) (1997) *Forest Decline and Ozone. A Comparison of Controlled Chamber and Field Experiments*, Ecological Studies, Vol. 127. Springer-Verlag, Berlin.

Savchenko, V.K. (1995) *The Ecology of the Chernobyl Catastrophe: Scientific Outlines of an International Programme of Collaborative Research* (Man and Biosphere Series, Vol. 16). UNESCO, Paris, and Parthenon Publishing Group, Carnforth, UK.

Schulze, E.-D. (1989) Air pollution and forest decline in a spruce (*Picea abies*) forest. *Science* 244, 776–783.

Schwarz, O. (1997) Management of forest ecosystems in the Karkonose National Park, Black Triangle region, Czech Republic. In: Gutkowski, R.M. and Winnicki, T. (eds) *Restoration of Forests. Environmental Challenges in Central and Eastern Europe*. Kluwer Academic, Dordrecht, pp. 215–226.

Shcheglov, A.I., Tsvetnova, O.B. and Tikhomirov, F.A. (1992) Migration of long-living radionuclides of Chernobyl fallout in the forest soils of the European part of the Commonwealth of independent states. *Moscow University Soil Science Bulletin* 47, 23–30.

Shubin, V.A. (1996) Forestry in the Chernobyl zone of Russia. *Lesnoe Khoz* 5 (1996), 2–5 (in Russian).

Skuterud, L., Travnikova, I.G., Balonov, M.I., Strand, P. and Howard, B.J. (1997) Contribution of fungi to radiocaesium intake by rural populations in Russia. *Science of the Total Environment* 193, 237–242.

Slovik, S., Siegmund, A., Führer, H.-W. and Heber, U. (1996) Stomatal uptake of SO_2, NO_x and O_3 by spruce crowns (*Picea abies*) and canopy damage in Central Europe. *New Phytologist* 132, 661–676.

Solberg, S. and Torseth, K. (1997) Crown condition of Norway spruce in relation to sulphur and nitrogen deposition and soil properties in southeast Norway. *Environmental Pollution* 96, 19–27.

Spiecker, H., Mielikäinen, K., Köhl, M. and Skovsgaard, J.P. (eds) (1996) *Growth Trends in European Forests*. Springer-Verlag, Berlin.

Stachurski, A., Zimka, J.R. and Kwiecien, M. (1995) Forest decline in Karkonosze (Poland). I. Chlorophyll, phenols, defoliation index and nutrient status of the Norway spruce (*Picea abies* L.). *Ekologia Polski* 42, 289–316.

Taylor, G.E. Jr, Johnson, D.W. and Andersen, C.P. (1994) Air pollution and forest ecosystems: a regional to global perspective. *Ecological Applications* 44, 662–689.

Tikhomirov, F.A., Shcheglov, A.I., Tsvetnova, O.B. and Klyashtorin, A.L. (1991) Geochemical migration of radionuclides in forest ecosystems within the radioactive contamination zone of the Chernobyl Nuclear Power Plant. *Soviet Soil Science* 23, 66–75.

Tikhomirov, F.A., Shcheglov, A.I. and Tsvetnova, O.B. (1993a) Radiation protection measures in forests within the radioactive exclusion zone of the Chernobyl nuclear power station, and evaluation of their effectiveness. *Lesnoe Khoz* 4 (1993), 30–32 (in Russian).

Tikhomirov, F.A., Sheheglov, A.I. and Sidorov, V.P. (1993b) Forests and forestry: radiation protection measures with special reference to the Chernobyl accident zone. *Science of the Total Environment* 137, 289–305.

Tikhomirov, F.A., Sheheglov, A.I. and Sidorov, V.P. (1994) Forests and forestry in conditions of radioactive pollution. *Lesnoe Khoz* 1, 26–29 (in Russian).

Tjoelker, M.G., Volin, J.C., Oleksyn, J. and Reich, P.B. (1995) Interaction of ozone pollution and light on photosynthesis in a forest canopy experiment. *Plant, Cell and Environment* 18, 895–905.

Ulrich, B. (1993) Interaction of forest canopies with atmospheric constituents: SO_2, alkali and earth alkali cations and chloride. In: Ulrich, B. and Pankrath, J. (eds) *Effects of Accumulation of Air Pollutants in Forest Ecosystems*. D. Riedel, Dordrecht, pp. 34–45.

Ulrich, B., Mayer, R. and Khanna, P.K. (1980) Chemical changes due to acid precipitation in a loss-derived soil in central Europe. *Soil Science* 130, 193–199.

Ushakov, B.A., Panfilov, A.V. and Vasilenko, A.A. (1992) Radioactive pollution of the forests of the Bryansk region. *Lesnoe Khoz* 1, 29–30 (in Russian).

Vancura, K. (1992) Forestry in Air Polluted Areas. UN/ECE workshop and study tour through air polluted regions of Czechoslovakia. V/LHM 5/1992.

Vasicek, J. (1997) *Report on Forestry of the Czech Republic 1996*. Publications of the Ministry of Agriculture of the Czech Republic, Praha.

Wiczkiewicz, M. (1982) Historical outlook of forest management in the Sudety Mountains. *Sylwan* 6, 49–54 (in Polish).

Yushkov, P.I., Chueva, T.A. and Kulikov, N.V. (1994) Effect of increased background of ionizing radiation on European birch in the zone of the Chernobyl accident. *Russian Journal of Ecology* 24, 309–314.

Zahradnik, P. (1997) *Forest Protection Newsletter*, publications of Forestry and Game Management Research Institute. Forest Protection Service, Strnady.

Zeller, K., Cerny, M., Bytnerowicz, A., Smith, L., Sestak, M., Michalec, M., Pernegr, V. and Kucera, J. (1997) Air pollution status of a representative site in the Czech Republic Brdy Mountains. *Environmental Pollution* 98, 291–297.

Ozone Impacts on Californian Forests

9

P.R. Miller and M.J. Arbaugh

Pacific Southwest Research Station, Forest Service, USDA, Riverside, California, USA

The history of ozone effects on the forests of southern California can be traced by records of crown condition or stem growth. For example, during the period from 1945 to 1972 the growth increment of *Pinus ponderosa* Dougl. Ex P. & C. Laws, near the town of Blue Jay in the San Bernardino Mountains decreased every year and then began to recover to pre-1945 levels between 1973 and 1991. The growth recovery was coincident with an improvement of ozone air quality in the nearby South Coast Air Basin of California. Ozone injury on ponderosa pine in the southern Sierra Nevada was first discovered in the early 1970s. Despite general improvements of air quality in upwind urban regions, crown injury to sensitive ponderosa pines was detectable and measurable at numerous locations in California mountains by the mid-1990s. This paper describes the methods for evaluating ozone injury and interpreting effects at the individual tree and ecosystem levels.

1 Introduction

In California, ozone-induced injury to trees depends on the terrain or topographic position, climate and urbanization. Topography is important because pollution accumulates in valleys or air basins near contiguous mountains and high elevation areas that form the boundaries of air basins and support native conifer forests. The best examples are the San Gabriel and San Bernardino mountains (the Angeles and San Bernardino National Forests) of southern

California, which form the northern and eastern boundaries of the south coast air basin, and the central and southern Sierra Nevada (including numerous National Forests and two National Parks) which form the eastern side of the San Joaquin air basin (Fig. 9.1). The basin terrain supports intense urbanization and numerous stationary and mobile sources of hydrocarbons and NO_x, the precursors of tropospheric ozone. The summer season climate is dominated by large-scale high pressure systems over the south-western United States which generate subsidence inversions that confine pollutants to the air basins for many hours each day. In the afternoon, local sea breezes and upslope winds tend to transport pollutants to inland mountains as inversion layers are temporarily disrupted. Ozone transported from the urban source areas to distant forested mountains is a colourless gas mixed with the grey, visibility-obscuring haze, commonly known as photochemical smog.

Fig. 9.1. California map showing National Forests and National Parks in the Sierra Nevada and National Forests in southern California exposed to ozone transported from major urban centres.

Tree ring analysis has provided a record of the temporal influence of both ozone exposure and drought. For example, the annual basal area increment of a group of ponderosa pines at Blue Jay, California, between 1880 and 1990 illustrates (Fig. 9.2) responses that are characterized by three distinct time periods. The first period from 1880 to 1945 is represented by increasing basal area increments. The second period from 1946 to 1972 is characterized by declining basal area increments (by more than 40%). From 1973 to 1991 basal area growth rates returned to levels similar to those prior to 1945 (Arbaugh *et al.*, 1999a). Measures of ponderosa and Jeffrey pine (*Pinus jeffreyi* Grev. & Balf.) crown condition across the ozone gradient in the San Bernardino Mountains between 1973 and 1988 using the forest pest management (FPM) score, and the ozone injury index (OII) in 1995, show a distinct improvement of crown health at all but one heavily exposed sample plot, Camp Paivika (CP), located in the western region of the San Bernardino National Forest. The ozone injury score also showed this improvement between 1974 and 1988 (Miller *et al.*, 1989), which is coincident with a decrease in ozone exposure (Miller and McBride, 1989). This description generally confirms that ponderosa and Jeffrey pine health has improved to nearly pre-pollution levels in the more distant regions of the San Bernardino Mountains. However, studies conducted in the early 1990s indicate that crown injury varies from year to year, depending on moisture availability to trees and ozone uptake. There is thus potential for increased injury as long- and short-term climate patterns change.

In the Sierra Nevada the most recent survey data show continuing crown injury increasing from north to south in that mountain range (Arbaugh *et al.*,

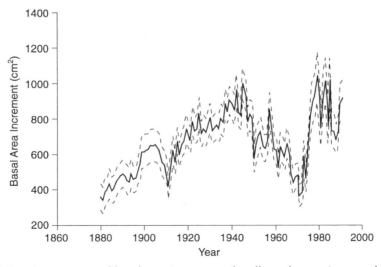

Fig. 9.2. Average annual basal area increments for all ponderosa pine sampled at Blue Jay, California. Dashed lines re-approximate ± 95% confidence intervals. Sample size varied through time (*n* = 40 at 1880, *n* = 166 at 1991).

1999b). The ozone exposure–crown injury relationship has been described for trees in the Sierra Nevada transect (Fig. 9.3) during the 1991–1995 period. The circumstances of tree injury in the Sierra Nevada are described in greater detail later in this paper.

2 Survey and sampling methods

The first method to test evidence that ozone was the cause of foliar symptoms on ponderosa pine employed branchlets of understorey ponderosa pines contained in three branch chamber treatments including, carbon-filtered air alone, 0.5 ppm ozone added to filtered air and ambient air (Miller *et al.*, 1963). Visible chlorotic mottle symptoms and needle abscission increased in the ozone-enhanced and ambient air treatments and diminished in the carbon-filtered air treatment. Chlorophyll extractions showed increased chlorophyll concentrations in the filtered-air treatment and reduced concentrations in the ambient air and the ozone-added treatments. Ozone fumigation experiments with ponderosa pine seedlings by Richards *et al.* (1968) confirmed that ozone caused the chlorotic mottle symptom. The presence or absence of this symptom has been the primary visual symptom for use in subsequent years for effects assessment in field surveys and fumigation experiments.

The relationship of chlorotic mottle to 38 other measures including morphological, physiological or nearest-neighbour response on the same trees

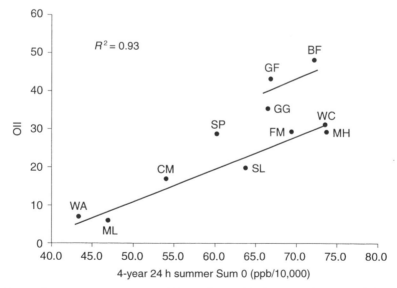

Fig. 9.3.   Ozone exposure–crown injury relationship between Sum 0 and OII. The upper line represents three sites that had > 90% of trees injured, and the lower line is from sites that had < 90% of trees injured.

were compared to determine the minimum number of these characteristics which contributed to ozone injury as recognized by chlorotic mottle (Grulke and Lee, 1997). Nine characteristics which had the highest correlations with differing amounts of chlorotic mottle were:

1. The number of green whorls retained;
2. The proportion of foliated versus total branchlet length;
3. Branchlet diameter;
4. Bole diameter;
5. Radial growth;
6. Foliar chlorophyll content;
7. Foliar nitrogen content;
8. Coarse root sucrose content; and
9. The proximity of the nearest *Pinus* neighbour.

Of these nine variables there were five that were superior, namely, 1, 2, (3 or 4), 6 and 7. Morphological characteristics had a greater influence in classifying chlorotic mottle than did physiological or nearest-neighbour characteristics. Without chlorotic mottle to signify the presence of ozone injury, any one of the above characteristics could be attributed to a number of possible stressors. Several studies have repeatedly implicated visual 'chlorotic mottle' as a specific pathological symptom that is attributable to ozone (Miller *et al.*, 1963; Richards *et al.*, 1968; Temple *et al.*, 1992). Mesophyll cell degradation has also been associated with chlorotic mottle (Evans and Miller, 1972).

Chlorotic mottle presence or absence or amount has been the key element of field survey methods leading to the expression of the amount of ozone injury as a score or index. The three indices of crown injury that have been employed are referred to as the FPM index (Pronos *et al.*, 1978), the ozone injury score (OIS) (Miller, 1973), and the OII (Duriscoe, 1988; Miller *et al.*, 1996a). The FPM method has one major feature in common with the other procedures, namely, a determination of the youngest age cohort (whorl) of needles that have chlorotic mottle symptoms, the definitive symptom of ozone injury. Accordingly, the worst injury is rated as 0 and the least was set at 5, assuming that the five youngest whorls are more productive and valuable to retain. Recent analyses (Arbaugh *et al.*, 1998) indicate that all three methods yield similar information, when the means of 50 trees at a site are compared.

Presently the method most favoured for evaluating ozone injury to the crowns of pines is the OII and the explanation of field sampling and data analysis methods have been included in a descriptive manual (Miller *et al.*, 1996a). One of the most reliable aspects of this method is that a five-branchlet sample is pruned from each tree and morphological features and mottle symptoms are measured or estimated in-hand under uniform natural lighting. Crews can be trained successfully to perform the evaluation within the tolerances of specified quality assurance objectives (Miller *et al.*, 1996a).

The measured or estimated features include number of annual needle whorls present, the percentage of needle fascicles remaining in each needle whorl (0–33%, 34–66% or 67–100%), the modal length of needles in each whorl and the percentage of needle surface area showing chlorotic mottle in each whorl (0 = 0%, 1 = 1–6%, 2 = 7–25%, 3 = 26–50%, 4 = 51–75%, 5 = 76–100%). The total branchlet foliated length is also measured. However, the algorithm for computing the ozone injury index does not include foliated length, but does include percentage live crown. In the final computation, chlorotic mottle and whorl retention are each weighted 40%, and needle length and percentage live crown are each weighted 10%. In theory, the OII ranges between 0 = no injury and 100 = maximum injury. In practice, the mean maximum index for areas experiencing maximum ozone exposure is presently around 65 (in the western portion of the San Bernardino Mountains) and further analysis results indicate that 75–85 may be the observable maximum. Chlorotic mottle must be present on at least one whorl before the index is computed. The method has been field tested since 1991 in the Forest Ozone Response Study located primarily in the Sierra Nevada and results are presented below (Arbaugh *et al.*, 1998).

Regressions of the indices from sample populations for FPM and OIS and for FPM and OII resulted in r^2 values ≥ 0.95 (Arbaugh *et al.*, 1998). This lends great confidence to the interchangeable use of these indices for estimating ozone injury. During the 1980s, in particular, there was intensive development, testing and application of methods for the visual evaluation of forest health in Europe and the United States (Innes, 1993a). Methods associated with the assessment of trees in forests have relied heavily on crown transparency (density) and crown discoloration (Innes, 1993b); unfortunately these are non-specific indicators when it comes to distinguishing the effects of air pollutants from other environmental stresses. In addition, these measures are subjective and susceptible to serious error. Fortunately, there has been general agreement in the western United States that chlorotic mottle is associated with ozone exposure, therefore it has been available to use as a specific effect of ozone on pine foliage, and its application has not met with the same uncertainties associated with the use of crown density and discoloration.

In California the evaluation of ozone injury to companion tree species has not met with the same level of acceptance and wide use as is the case with ponderosa or Jeffrey pine. Either the chlorotic mottle symptom does not develop distinctly prior to leaf abscission or the natural distribution of some species across a gradient of ozone exposure is limited compared with pine. At the beginning of the monitoring work in the San Bernardino Mountain plots attempts were made to evaluate ozone injury symptoms on all tree species in each plot in order to determine the most sensitive indicators of year-to-year changes in ozone injury. The general features of the OIS were applied to the description of ozone injury to sugar pine (*Pinus lambertiana* Dougl.), white fir (*Abies concolor* (Gord. & Glend.) Lindl. ex Hildbr.) and incense cedar (*Libocedrus decurrens* Torr.) present in the same plots with pines along the ozone gradient (Miller *et al.*, 1977). White fir appeared in 15 of 19 plots, sugar pine in 6 of 19 plots and incense cedar in 5 of 19 plots. The white fir scores for 1973–1975 were the lowest where ponderosa pine scores were also low and the highest where there was very slight or no injury to ponderosa or Jeffrey pines. Even though white fir did show a gradient of foliar injury similar to pines, it was only moderately sensitive compared to ponderosa or Jeffrey pines and was not present at every location along the gradient. Therefore, the use of white fir as a bioindicator was de-emphasized. Incense cedar and sugar pine did not show injury trends at the few locations where they were present and they did show advanced defoliation and a general yellowing of foliage, but without chlorotic mottle as seen on ponderosa pine and Jeffrey pine.

California black oak (*Quercus kelloggii* Newb.) was present on 15 of 19 plots. A visual rating scale was used to rate injury to leaves of the upper and lower crown, and the two estimates for each tree were added to produce an injury rating which ranged from 0 to 8 (Miller *et al.*, 1980). Oaks also showed a gradient of ozone injury which ranged from 4.8 at one of the highest ozone exposure plots to 8.0 at the lowest exposure plots. These ratings were an average of annual scores from 1975 to 1978 (Miller *et al.*, 1980). The oaks were less sensitive to ozone than ponderosa pine and Jeffrey pine, and the deciduous leaves were useful mostly to single out particular years or particular places with higher ozone exposure. These initial observations of companion tree species led to the decision to focus research and monitoring on ponderosa and Jeffrey pines as the two most ubiquitous and ozone-sensitive tree species.

More recent work has examined injury to bigcone Douglas fir (*Pseudotsuga macrocarpa* (Vasey) Mayr). This coniferous species is found only in the mountains of southern California and Baja California, Mexico, between elevations of 700 and 2200 m in canyons and dry slopes. A study of 18 sites of bigcone Douglas fir across an air pollution gradient was conducted in 1988 (Peterson *et al.*, 1995). Analysis of the foliage across this gradient indicated that needle retention was lower at high- and moderate-ozone sites than at low-ozone sites, despite the lack of chlorotic mottle symptoms on the foliage.

3   Present extent of ozone injury to ponderosa and Jeffrey pines

3.1   San Bernardino Mountains

The source area for pollutants transported to the conifer forests of the San Bernardino Mountains is the South Coast Air Basin of southern California. Between 1976 and 1991 the weather-adjusted ozone data for the May to October 'smog season' showed that the number of days exceeding the Federal Standard (> 120 ppb, 1 h average) declined at an average annual rate

of 2.27 days per year (Davidson, 1993) for this air basin. The number of days
with Stage I episodes (> 200 ppb, 1 h average) declined at an average annual
rate of 4.7 days per year during the same time period. The total days per year
with exceedances of the Federal Standard was as high as 159 in 1978 and the
lowest was 105 days in 1990. During Stage I episodes the high was 108 days
in 1979 and the low was 33 days in 1990. The 1974 to 1988 trends of the
May to October hourly average and the average of monthly maximum ozone
concentrations for Lake Gregory, a forested area in the western section of the
San Bernardino Mountains, also showed a decline (Miller and McBride, 1989).
This declining trend has continued through 1997, which had the lowest
pollution levels ever recorded for this air basin.

The ozone injury score has also shown an improvement in chronic injury
to crowns of ponderosa and Jeffrey pines between 1974 and 1988, in 13 of 15
plots located on the gradient of decreasing ozone exposure in the San
Bernardino Mountains (Miller and McBride, 1989). The two exceptions were
plots located at the highest exposure end of the gradient which also receive
large amounts of nitrogenous deposition (Fenn and Bytnerowicz, 1993).

Miller *et al.* (1989) reported that for the 1974–1988 period the basal area
increase of ponderosa pines was generally less than competing species at 12 of
the 13 plots evaluated. The total basal area for each species, as a percentage
of the total basal area for all species, shows that ponderosa and Jeffrey pines lost
basal area in relation to competing species that were more tolerant to ozone,
namely, white fir, incense cedar, sugar pine and black oak, at plots with slight
to severe crown injury to ponderosa or Jeffrey pine. The accumulation of more
stems of ozone-tolerant species in the understorey presents a fuel ladder
situation that jeopardizes the remaining overstorey trees in the event of a catastrophic fire. The ozone-tolerant species are inherently more susceptible to fire
damage because of thinner bark and branches close to the ground (Minnich,
1999).

Annual changes in crown condition between 1991 and 1994 illustrate
that ozone exposure is still a significant influence on tree health at present-day
concentrations and durations. The usefulness of the OII for tracking year-to-
year changes was tested in annual evaluations from 1991 to 1994 of 130
ponderosa or Jeffrey pines distributed in three plots in the Barton Flats area, a
location in the mid-range of the ozone exposure gradient across the San
Bernardino Mountains (Miller *et al.*, 1998). Over the course of the 1991 to
1994 study, ozone continued to cause injury each year. The frequency
distributions of the 130 sampled ponderosa or Jeffrey pines shows that a shift
towards higher numbers in higher injury ranges began in 1992 and remained
the same in 1993 and 1994 (Fig. 9.4). The corresponding OII means for all
sample trees were: 1991, 38.3; 1992, 47.5; 1993, 47.5; and 1994, 49.2. The
only significant difference between years ($P < 0.05$), detected with a paired
T-test, was between 1991 and 1992. This change was related to a higher level
of ozone flux to foliage in 1992, the first year with favourable soil moisture

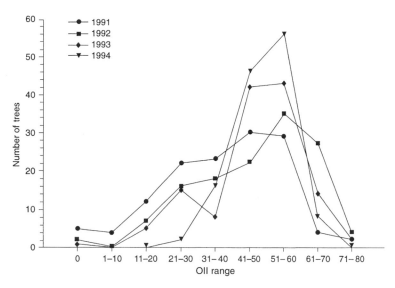

Fig. 9.4. Annual changes of the distribution in 10 increment intervals of the ozone injury index (OII) of 130 sample ponderosa or Jeffrey pines near Barton Flats, California, in 1991 to 1994. Beginning in 1992 there was a shift to larger numbers of trees in higher injury intervals due to a larger flux of ozone to foliage in the early summer of 1992.

following several preceding years of drought (Miller *et al.*, 1998). Readily available soil moisture increases stomatal conductance, which results in greater gas exchange, i.e. ozone uptake by foliage. These results, from an area with moderate ozone exposure, confirm that ozone injury has remained fairly constant from year to year in recent years with the exception of 1992, when it was higher.

3.2 Western slope of the Sierra Nevada mountains

Detection of ozone injury symptoms to ponderosa and Jeffrey pines in the Sierra Nevada, California (Miller and Millecan, 1971) and subsequent surveys by Forest Service pest management specialists using 10-tree trend plots (Pronos and Vogler, 1981), provided the earliest data describing the extent of ozone injury and the early trends of the severity of injury. For example, Pronos and Vogler (1981) reported that between 1977 and 1980 the general trend was for an increase in the extent of ozone symptoms present on pine foliage.

Peterson and others (1991) sampled crown condition and derived basal area growth trends from cores collected from ponderosa pines at sites in seven

Federal administrative units (National Forests and National Parks) located from north to south in the Sierra Nevada including Tahoe National Forest, Eldorado National Forest, Stanislaus National Forest, Yosemite National Park, Sierra National Forest, Sequoia-Kings Canyon National Park, and Sequoia National Forest. In July–August 1987, four symptomatic and four asymptomatic sites were visited in each unit and only sites with ponderosa pines greater than 50 years old were selected for sampling. The symptomatic plots generally indicated increasing levels of chronic ozone injury (reduced numbers of annual needle whorls retained and chlorotic mottle symptoms on younger age classes of needles) from north to south. In general, the results of this study documented the regional nature of the ozone pollution problem originating primarily from the San Joaquin Valley Air Basin, as well as the San Francisco Bay Air Basin further to the west. The study found no evidence of recent large-scale growth changes in ponderosa pine growth in the Sierra Nevada mountains; however, the frequency of trees with recent declines of growth did increase in the southernmost units. Because these units had the highest levels of ozone (and more chlorotic mottle symptoms on needles of younger age classes), it is likely that ozone is one of the factors contributing to decline in basal area growth. Other factors limiting tree growth in this region include periodic drought, brush competition and high levels of tree stocking.

This region-wide survey (Peterson *et al.*, 1991) of ponderosa pine provides a useful context for a number of other more narrowly focused studies or surveys in the Sierra Nevada. Another tree ring analysis and crown injury study was focused on Jeffrey pines in Sequoia-Kings Canyon National Park (Peterson *et al.*, 1987). This study suggested that decreases of radial growth of large, dominant Jeffrey pines growing on xeric sites (thin soils, low moisture-holding capacity) and exposed to direct upslope transport of ozone, resulted in as much as 11% less growth in recent years compared with similar trees protected from ozone exposure.

Both permanent plots and cruise surveys have been employed in Sequoia, Kings Canyon and Yosemite National Parks to determine the spatial distribution and temporal changes of injury to ponderosa and Jeffrey pine within these National Parks (Duriscoe and Stolte, 1989). Comparisons of the same trees at 28 plots between 1980–1982 and 1984–1985 in Sequoia National Park showed increased ozone injury for many trees and increased numbers of trees with ozone injury. Ozone injury was found to decrease with increasing elevation of plots. The highest levels of tree injury in the Marble Fork drainage of the Kaweah River were at approximately 1800 m elevation and were associated with hourly peak averages of ozone of 80–100 ppb, but seldom exceeding 120 ppb.

A cruise survey in 1986 evaluated 3120 ponderosa or Jeffrey pines in Sequoia National Park and Yosemite National Park for ozone injury (Duriscoe and Stolte, 1989). More than one-third of these trees were found to have some level of chlorotic mottle. At Sequoia National Park, 39% of the sample comprised symptomatic trees (574 out of 1470) and at Yosemite National Park 29% (479 out of 1650) of the sample. Ponderosa pines were generally more severely injured than Jeffrey pines. The FPM score (low score equals high injury) was 3.09 for ponderosa pines and 3.62 for Jeffrey pines (Pronos *et al.*, 1978). These cruise surveys also identified the spatial distribution of injury in Sequoia and Yosemite National Parks, which indicated that trees in the drainages nearest the San Joaquin Valley were the most injured.

The Lake Tahoe Basin is located at the northern end of the Sierra Nevada sampling transect (near the Eldorado National Forest) (Peterson *et al.*, 1991). The basin is distinct because its air quality is a combination of local and remote pollution sources, in contrast to most other Sierra Nevada sites where pollution results from long-range transport only. A survey in 1987 of 360 trees in 24 randomly selected plots in the basin found that 105 (29.2%) had some level of foliar injury (Pedersen, 1989). Seventeen of these plots had FPM injury scores (Pronos *et al.*, 1978) that fell in the slight injury category. Sixteen additional cruise plots containing 190 trees extended observations to the east outside the basin found 21.6% of trees had foliar injury, less than in the basin.

Since about 1992, the Forest Ozone REsponse STudy (FOREST) has monitored the condition of pines and ozone air quality at ten locations on a north to south transect in the Sierra Nevada from Lassen Volcanic National Park in the north to a location south of Sequoia National Park in the south. One additional site where crown injury is moderate was also included further south in the San Bernardino Mountains. The Sierra locations ranged from almost no crown injury (near zero OII) due to ozone in the north and gradually increased to moderate crown injury in the south. The single site where crown injury is moderate in the San Bernardino Mountains is located about midway along a west to east gradient of ozone exposure. Tree response has been ana- lysed in the FOREST study in relation to several ozone exposure indices from the nearest monitoring site (Arbaugh *et al.*, 1998). Significant associations were found between OII and 4-year 24-h summer SUM0, SUM06, W126 and HRS80 (number of hours exceeding 80 pphm) ozone indices. These statistical associations are not adjusted for the influence of variable seasonal ozone flux to foliage, which differs each year depending on soil moisture availability (Temple and Miller, 1999).

Giant sequoia seedlings

In Sequoia National Park, field plot observations of seedling health and mortal- ity in natural giant sequoia groves were studied from 1983 to 1986. Results showed that emergent seedlings in moist microhabitats had ozone-induced foliar symptoms. High seedling mortality from drought and other abiotic factors also occurred during this period. A variable such as ozone, which could injure seedling foliage sufficiently to reduce root growth immediately after germination, could increase vulnerability to late summer drought. After fumigation, giant sequoia (*Sequoiadendron giganteum* Bucch.) seedlings developed chlorotic mottle following *in situ* exposure to both ambient ozone concentrations and 1.5×ambient ozone in open-top chambers during the 8–10 weeks after germination (Miller *et al.*, 1994). Significant differences in light compensation point, net assimilation at light saturation and dark respiration were found between seedlings in charcoal filtered air treatments and 1.5×ambient ozone treatments (Grulke *et al.*, 1989). These results indicate that ozone has the potential to be a new selection pressure during the regeneration phase of giant sequoia, possibly leading to reduced genetic diversity.

4   The role of other abiotic and biotic stresses in the forest ecosystem

The knowledge attained over the years has provided the first ecosystem level interpretation of chronic ozone exposure effects on a forest ecosystem in the USEPA Criteria Documents, used in support of the evaluation of National Air Quality Standards for ozone. The extension of these observations into the future is clearly an advantage because few forest ecosystems in the West have quality-assured data sets with this length of record. Broad interpretations of ecosystem level responses have been published as a result of work completed in the 1970s to the 1990s (Miller *et al.*, 1977; Miller *et al.*, 1982; Miller *et al.*, 1996b; Miller *et al.*, 1998). Not as much attention has been devoted to the interpretation of the Sierra Nevada ecosystem-level research, although the pine and mixed conifer forest types are analogous to those occurring in the San Bernardino Mountains.

   One set of relationships that deserves discussion is that of pine bark beetles (*Coleoptera*: *Scolytidae*), drought and tree mortality. Stark *et al.* (1968) confirmed the relationship between chronic ozone injury and increased incidence of bark beetle attack. In the preceding decade foresters in the San Bernardino Mountains had tried to cope with the heavy incidence of bark beetle infestation in the Barton Flats area of the San Bernardino Mountains by salvage logging of weakened trees (Hall, 1958). Drought can seriously accelerate the success of bark beetle attacks on pines. We have only a vague understanding of the drought, bark beetle and tree mortality complex. Certainly the two stressors acting simultaneously would be additive in their influence on bark beetle-caused mortality; however, sequential action could also be important. Low precipitation years (droughts) frequently occur as clusters of 2–5 years' duration. The uptake of ozone during these years is diminished because of lower stomatal conductance due to lower summer season soil moisture levels. During years of higher precipitation, conductance and ozone uptake increase and foliar injury symptoms increase (Temple and Miller, 1999). This problem is aggravated by short needle lengths grown during the drought years, so that the total photosynthetic surface area is lower than in pre-drought years. Thus, increased ozone injury during the time that trees are recovering from drought stress may increase tree vulnerability to bark beetle attacks.

Another important issue is the role of both dry and wet deposited nitrogen. Several forms of atmospheric nitrogen, especially from dry deposition, are an important part of the photochemical pollutant mixture containing ozone. Nitrogen accumulates and can saturate the litter layer and surface soil. One theory is that nitrogen may help trees to compensate for ozone injury. Preliminary observations suggest that bole growth is not inhibited in areas with high nitrogen deposition even though crown injury from ozone remains high. Results from scenarios simulated by a carbon flow model, Century (Parton *et al.*, 1987), indicate that increased ozone and nitrogen leads to increased foliar production and litter biomass (Fig. 9.5) (Arbaugh *et al.*, 1999b). Measurements of litter depth at high ozone sites support this result (Minnich *et al.*, 1995; Minnich, 1999). Continued nitrogen input may lead to long-term changes in soil chemistry. Simulations using the nutrient cycling model (NuCM) predicted that increased N deposition will result in N saturation and groundwater pollution by NO_3^-, and possibly to K^+ deficiency (Miller *et al.*, 1996b; Fenn *et al.*, 1999). An experiment is now underway that is testing the effects of the repeated application of nitrogen fertilizer in forest stands subject to either low or high ozone exposure.

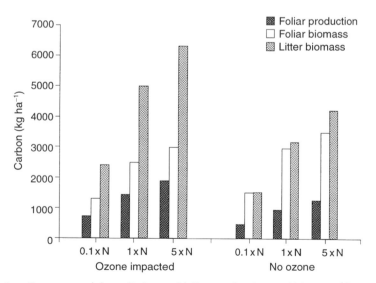

Fig. 9.5. Century model predictions of foliar production and biomass changes associated with doubled foliar turnover rates at different levels of N deposition. Notice that as N deposition increases, the effect of ozone-N deposition on litter biomass increases, indicating that the effect of the two pollutants together may be interactive rather than additive.

5 Conclusions

The short history of ozone injury to Californian coniferous forests shows that the initial years of severe exposure, which resulted in substantial crown injury and increased mortality, have stabilized in recent years at lower levels of crown injury which are easily distinguishable by visual detection methods. Concern has been raised about the vulnerability of these forest stands to wildfire, because stand composition is shifting towards higher proportions of those species (mostly ozone-tolerant species) that present a green fuel ladder and are more heat sensitive because of thin bark (this includes typically incense cedar or white fir), and because of the increasing depth of the litter layer. The open spaces left by dying pines (during the 1970s) and the higher reproductive capacity of the remaining incense cedar and white fir have combined to increase forest density. Nitrogen deposition has also contributed to increased stand density by increasing tree growth rates. Current levels of ozone exposure and periodic drought stress will continue to threaten the vigour of ponderosa and Jeffrey pines directly and indirectly through their reduced competitive ability with other tree species, decreased resistance to bark beetle attacks and higher risk of stand-replacing fires.

Acknowledgements

Thanks are extended to the agencies who have shared in the support of research in the San Bernardino Mountains and Sierra Nevada Mountains over many years, including the US Environmental Protection Agency, the California Air Resources Board and the National Park Service.

References

Arbaugh, M.J., Miller, P.R., Carroll, J., Takemoto, B. and Procter, T. (1998) Relationship of ambient ozone with injury to pines in the Sierra Nevada and San Bernardino mountains of California, USA. *Environmental Pollution* 101, 291–301.

Arbaugh, M.J., Peterson, D.L. and Miller, P.R. (1999a) Air pollution effects on growth of Ponderosa pine, Jeffrey pine, and Bigcone Douglas-fir. In: Miller, P.R. and McBride, J.R. (eds) *Oxidant Air Pollution Impacts in the Montane Forests of Southern California. A Case Study of the San Bernardino Mountains.* Springer Verlag, New York, pp. 179–207.

Arbaugh, M.J., Johnson, D.W. and Pulliam, W. (1999b) Simulated effects of N deposition, ozone injury, and climate change on a forest stand in the San Bernardino Mountains. In: Miller, P.R. and McBride, J.R. (eds) *Oxidant Air Pollution Impacts in the Montane Forests of Southern California. A Case Study of the San Bernardino Mountains.* Springer Verlag, New York, pp. 353–372.

Davidson, A. (1993) Update on ozone trends in California's south coast air basin. *Journal of the Air Pollution Control Association* 31, 38–41.

Duriscoe, D.M. (1988) *Methods of Sampling for* Pinus ponderosa *and* Pinus jeffreyi *for the Evaluation of Oxidant Induced Foliar Injury.* Final Report, Eridanus Research Associates, Three Rivers, California.

Duriscoe, D.M. and Stolte, K.W. (1989) Photochemical oxidant injury to ponderosa (*Pinus ponderosa* Dougl. ex Laws) and Jeffrey pine (*Pinus jeffreyi* Grev. and Balf.) in the national parks of the Sierra Nevada of California. In: Olson, R.K. and Lefohn, A.S. (eds) *Effects of Air Pollution on Western Forests.* Transactions Series, No.16, Air and Waste Management Association, Pittsburgh, pp. 261–278.

Evans, L.S. and Miller, P.R. (1972) Ozone damage to ponderosa pine – a histological and histochemical appraisal. *American Journal of Botany* 59, 297–304.

Fenn, M.E. and Bytnerowicz, A. (1993) Dry deposition of nitrogen and sulfur in the San Bernardino National Forest in southern California. *Environmental Pollution* 81, 277–285.

Fenn, M.E., Poth, M.A., Aber, J.D., Baron, J.S., Bormann, B.T., Johnson, D.W., Lemly, A.D., McNulty, S.G., Ryan, D.F. and Stottlemyer, R. (1999) Nitrogen excess in North American ecosystems: predisposing factors, ecosystem responses, and management strategies. *Ecological Applications* 8, 706–733.

Grulke, N.E. and Lee, E.H. (1997) Assessing visible ozone-induced foliar injury to ponderosa pine. *Canadian Journal of Forest Research* 27, 1658–1668.

Grulke, N.E., Miller, P.R., Wilborn, R.D. and Hahn, S. (1989) Photosynthetic response of giant sequoia seedlings and rooted branchlets of mature foliage to ozone fumigation. In: Olson, R.K. and Lefohn, A.S. (eds) *Effects of Air Pollution on Western Forests.* Transactions Series, No.16, Air and Waste Management Association, Pittsburgh, pp. 261–278.

Hall, R.C. (1958) Sanitation-salvage controls bark beetles in southern California recreation area. *Journal of Forestry* 56, 9–11.

Innes, J.L. (1993a) *Forest Health: Its Assessment and Status.* CAB International, Wallingford.

Innes, J.L. (1993b) Methods to estimate forest health. *Silva Fennica* 27, 145–157.

Miller, P.R. (1973) Oxidant-induced community change in a mixed conifer forest. In: *Air Pollution Damage to Vegetation.* Advances in Chemistry Series 122, pp. 101–117.

Miller, P.R. and McBride, J.R. (1989) Trends of ozone damage to conifer forests in the western United States, particularly southern California. In: Bucher, J.B. and Bucher-Wallin, I. (eds) *Proceedings of the 14th International Meeting for Specialists in Air Pollution Effects on Forest Ecosystems. IUFRO P2.05, Interlaken, Switzerland, 2–8 Oct 1988.* Swiss Federal Institute for Forest, Snow and Landscape Research, Birmensdorf, pp. 61–68.

Miller, P.R. and Millecan, A.A. (1971) Extent of oxidant air pollution damage to some pine and other conifers in California. *Plant Disease Reporter* 55, 555–559.

Miller, P.R., Parmeter, J.R., Taylor, O.C. and Cardiff, E.A. (1963) Ozone injury to the foliage of *Pinus ponderosa. Phytopathology* 53, 1072–1076.

Miller, P.R., Kickert, R.N., Taylor, O.C., Arkely, R.J., Cobb, F.W., Jr, Dahlsten, D.L., Gersper, P.J., Luck, J.R., McBride, J.R., Parmeter, J.R., Jr, Wenz, J.M., White, M. and Wilcox, W.W. (1977) *Photochemical Oxidant Air Pollutant Effects on a Mixed Conifer Forest Ecosystem – a Progress Report, 1975–1976.* Report EPA 600/3-77-104, Environmental Protection Agency, Corvallis.

Miller, P.R., Longbotham, G.J., Van Doren, R.E. and Thomas, M.A. (1980) Effect of chronic oxidant air pollution exposure on California black oak in the San Bernardino Mountains. In: *Proceedings of the Symposium on Ecology Management and Utilization of California Oaks, Claremont, California, 26–28 June 1979.* General Technical Report PSW 44, Pacific Southwest Forest and Range Experiment Station, Berkeley.

Miller, P.R., Taylor, O.C. and Wilhour, R.G. (1982) *Oxidant Air Pollution Effects on a Western Coniferous Forest Ecosystem.* Environmental Protection Agency, Research Brief, EPA-600/D-82-276, Environmental Protection Agency, Washington, DC.

Miller, P.R., McBride, J.R., Schilling, S.L. and Gomez, A.P. (1989) Trend of ozone damage to conifer forests between 1974 and 1988 in the San Bernardino Mountains of southern California. In: Olson, R.K. and Lefohn, A.S. (eds) *Effects of Air Pollution on Western Forests.* Air and Waste Management Association, Pittsburgh, pp. 309–323.

Miller, P.R., Grulke, N.E. and Stolte, K.W. (1994) Effects of air pollution on giant sequoia ecosystems. In: Aune, P.S. (tech. coord.) *Proceedings of the Symposium on Giant Sequoias: Their Place in the Ecosystem and Society, 23–25 June 1992, Visalia, California.* General Technical Report PSW-GTR-151, USDA, Forest Service, Pacific Southwest Research Station, Albany, California, pp. 90–98.

Miller, P.R., Stolte, K.W., Duriscoe, D. and Pronos, J. (Technical Coordinators) (1996a) *Monitoring Ozone Air Pollution Effects on Western Pine Forests.* General Technical Report 155, Pacific Southwest Research Station, Forest Service, US Department of Agriculture, Albany, California.

Miller, P.R., Poth, M.A., Bytnerowicz, A., Fenn, M.E., Temple, P.J., Chow, J., Watson, J., Frazier, C., Green, M. and Johnson, D. (1996b) *Ecosystem Level Alterations in Soil Nutrient Cycling: an Integrated Measure of Cumulative Effects of Acidic Deposition on a Mixed Conifer Forest in Southern California.* Final Report, Contract No.92-335, NTIS No. PB97-106-223, Environmental Protection Agency, Air Resources Board, Research Division, Sacramento, California.

Miller, P.R., Bytnerowicz, A., Fenn, M., Poth, M., Temple, P., Schilling, S., Jones, D., Johnson, D., Chow, J. and Watson, J. (1998) Multidisciplinary study of ozone, acidic deposition and climate effects on a mixed conifer forest in California, USA. *Chemosphere* 36, 1001–1006.

Minnich, R.A. (1999) Vegetation, fire regimes and forest dynamics. In: Miller, P.R. and McBride, J.R. (eds) *Oxidant Air Pollution Impacts in the Montane Forests of Southern California. A Case Study of the San Bernardino Mountains.* Springer Verlag, New York, pp. 44–80.

Minnich, R.A., Barbour, M.G., Burke, J.H. and Fernau, R.F. (1995) Sixty years of change in conifer forests of the San Bernardino Mountains: reconstruction of Californian mixed conifer forests prior to fire suppression. *Conservation Biology* 9, 902–914.

Parton, W.J., Schimel, D.S., Cole, C.V. and Ojima, D.S. (1987) Analysis of factors controlling soil organic levels of grasslands in the Great Plains. *Soil Science Society of America Journal* 51, 1173–1179.

Pedersen, B.S. (1989) Ozone injury Jeffrey and ponderosa pines surrounding Lake Tahoe, California and Nevada. In: Olson, R.K. and Lefohn, A.S. (eds) *Effects of Air Pollution on Western Forests.* Transactions Series, No.16, Air and Waste Management Association, Pittsburgh, pp. 279–292.

Peterson, D.L., Arbaugh, M.J., Wakefield, V.A. and Miller, P.R. (1987) Evidence of growth reduction in ozone-stressed Jeffrey pine (*Pinus jeffreyi* Grev. and Balf.) in Sequoia and Kings Canyon National Parks. *Journal of the Air Pollution Control Association* 37, 906–912.

Peterson, D.L., Arbaugh, M.J. and Robinson, L.J. (1991) Regional growth changes in ozone-stressed ponderosa pine (*Pinus ponderosa*) in the Sierra Nevada, California, USA. *Holocene* 1, 50–61.

Peterson, D.L., Silsbee, D.G., Poth, M.A., Arbaugh, M.J. and Biles, F.E. (1995) Growth responses of big cone Douglas fir (*Pseudotsuga macrocarpa*) to long term ozone exposure in southern California. *Journal of the Air and Waste Management Association* 45, 36–45.

Pronos, J. and Vogler, D.R. (1981) *Assessment of Ozone Injury to Pines in the Southern Sierra Nevada, 1979/1980*. Forest Pest Management Report 81–20, USDA Forest Service, Pacific Southwest Region, San Francisco.

Pronos, J., Vogler, D.R. and Smith, R.S. (1978) *An Evaluation of Ozone Injury to Pines in the Southern Sierra Nevada*. Forest Pest Management, Report 78–1, USDA Forest Service, Pacific Southwest Region, San Francisco.

Richards, B.L., Sr, Taylor, O.C. and Edmunds, G.F., Jr (1968) Ozone needle mottle of pine in southern California. *Journal of the Air Pollution Control Association* 18, 73–77.

Stark, R.W., Miller, P.R., Cobb, F.W., Jr, Wood, D.L. and Parmeter, J.R., Jr (1968) Photochemical oxidant injury and bark beetle (*Coleoptera: Scolytidae*) infestation of ponderosa pine. I. Incidence of bark beetle infestation in injured trees. *Hilgardia* 39, 121–126.

Temple P.J. and Miller, P.R. (1999) Seasonal Influences of ozone uptake and foliar injury to ponderosa and Jeffrey pines at a southern California site. In: Bytnerowicz, A., Arbaugh, M. and Schilling, S. (technical coordinators) *Proceedings of the International Symposium on Air Pollution and Climate Change Effects on Forest Ecosystems, 5–9 February 1996, Riverside, California*. US Department of Agriculture Forest Service GTR-166.

Temple, P.J., Riechers, G.H. and Miller, P.R. (1992) Foliar injury responses of ponderosa pine seedlings to ozone, wet and dry acidic deposition, and drought. *Environmental and Experimental Botany* 32, 101–113.

Ozone in the Mediterranean Region: Evidence of Injury to Vegetation

10

M.-J. Sanz and M.M. Millán

Centre for Environmental Studies of the Mediterranean, Valencia, Spain

Over the last 25 years, the Mediterranean basin has become heavily industrialized and, as a consequence, reports of air pollution damage to forested areas in southern European countries have increased since the late 1970s. Among pollutants, ozone appears to be one of the most potentially dangerous, both because the meteorological conditions present in the Mediterranean region are particularly favourable to its formation and persistence, and because it is the most ubiquitous. The available evidence strongly suggests that ozone occurs at sufficient concentrations to cause injury to sensitive plant species, including trees. Foliar injury caused by ambient ozone exposures has been clearly identified on native herbs, shrubs and tree species within several areas in southern Europe. However, ozone uptake and response are modulated by environmental factors such as drought. Recent studies suggest a complex interaction between high radiation, drought and ozone response in Mediterranean tree species such as Aleppo pine. More data are required before a full evaluation of the impact of ozone on Mediterranean forests can be achieved. Furthermore, additional research is needed to better understand and characterize the response of the Mediterranean ecosystems.

1   Introduction

Forest decline is considered to be a problem affecting the entire European continent. However, research on this topic in Mediterranean forests has not been a major feature of European research programmes, probably because 'acid rain' was never an important issue in the region. Over the last 25 years, however, the Mediterranean basin has become heavily industrialized and, as a consequence, reports of air pollution damage to forested areas in Spain, France and Greece have increased since the late 1970s (e.g. Naveh *et al.*, 1980; Velissariou *et al.*, 1992; Gimeno *et al.*, 1992; Garrec, 1994; Velissariou *et al.*, 1996; Sanz and Millán, 1998; Sanz and Calatayud, in press). The same is true of crops (Reinert *et al.*, 1992; Schenone and Lorenzini, 1992). Photooxidants are one of the most important pollutants in the Mediterranean area (Lorenzini *et al.*, 1984; Butkovic *et al.*, 1990; Millán *et al.*, 1992) because of the specific air pollution chemistry and dynamics of this region (Millán *et al.*, 1997). Recent meso-scale and regional models have reproduced observed photo-oxidant concentrations in the Mediterranean (Bastrup-Birk *et al.*, 1997; Thunis and Cuvelier, 1997), and indicate the presence of high concentrations in many non-monitored areas. Among the various pollutants, ozone appears to be one of the most potentially dangerous, both because the meteorological conditions of the Mediterranean region are particularly favourable to its formation and persistence, and because it is the most ubiquitous.

When studying the impact of air pollution on forests, knowledge of the meteorological processes across the entire Mediterranean in summer is very important for a proper interpretation of the data (Sanz and Millán, 1998). The dynamics of ozone in the region have been documented in several European Community (EC) supported projects (Millán *et al.*, 1992, 1996, 1997); these indicate that the horizontal distribution of air pollutants (i.e. ozone) in the Mediterranean basin is spatially non-homogeneous and that it undergoes seasonal variations and marked diurnal cycles in which meso-scale processes play an important role.

2   Ozone in the Mediterranean Basin

2.1   Ozone meso-scale dynamics in the Mediterranean region, with emphasis on the western basin

The EC has supported several large-scale projects to characterize the cycles of pollutants in this region, with emphasis on summer conditions, and to compose a mosaic of the atmospheric circulations involved. The MECAPIP project (1988–1991) documented the atmospheric processes over the Iberian peninsula. The RECAPMA project (1990–1991) extended the measurements from the Atlantic coast of Portugal to Italy, and the SECAP project (1992–1995) to the whole of the Mediterranean basin.

Meteorological processes across the entire Mediterranean in summer are dominated by two large, semi-permanent weather systems located at each end of the basin, as illustrated in Fig. 10.1. The Azores anticyclone lies at the

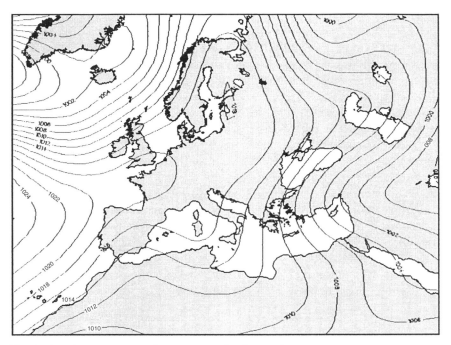

Fig. 10.1.   Average sea surface pressure for Europe in summer.

western edge, and the eastern borders have a low-pressure system which extends from the Middle East to south-western Asia (the monsoon system). In summer, the western Mediterranean basin is under strong insolation and weak levels of anticyclonic subsidence (decreasing from the coast of Portugal to Italy). These conditions favour the development of meso-scale flows and the recirculation of air masses (Millán *et al.*, 1992, 1996). During the same period the eastern Mediterranean basin is under conditions of weak ascent and strong advection, i.e. the Etesian winds, and the development of recirculations is largely inhibited (Millán *et al.*, 1997; Ziomas *et al.*, 1998).

The western Mediterranean is surrounded by high mountains, and on summer days their east- and south-facing slopes are strongly heated and act like orographic 'chimneys'. The 'chimneys' favour the early formation of upslope winds that reinforce the sea breezes and link the surface winds directly with their return flows aloft and their compensatory subsidence over the sea. This results in the formation of stacked layers along the coasts with the most recent layers at the top and older ones near the sea (Millán *et al.*, 1996). On the Spanish eastern coast, the layers reach 2–3 km in depth, have variable width over land (up to 100 km), and extend more than 300 km over the sea (Millán *et al.*, 1997).

These layers act as reservoirs for aged pollutants (ozone is considered to be a good indicator of aged air masses), and the lower ones can be brought inland by the sea breeze of the following day(s), creating recirculations, as summarized in Fig. 10.2. Tracer experiments on the Spanish eastern coast have shown that turnover times range from 2 to 3 days (Millán et al., 1992). During the night the land-based processes die out, and the reservoir layers may drift along the coasts and contribute to regional, interregional and long-range transport of aged pollutants. Similar processes have been documented in other regions of southern Europe (Ciccioli et al., 1987; Fortezza et al., 1993; Georgiadis et al., 1994; Orciari et al., 1999).

Under strong summer insolation, the coastal recirculations become 'large natural photochemical reactors' where most of the NO_x emissions and other precursors are transformed into oxidants, acidic compounds, aerosols and ozone. Relevant aspects of this problem are: (i) that the concept of downwind plumes is inappropriate in regions of complex coastal terrain where recirculation processes are important; (ii) that ozone is generated at the regional scale from emissions in urban centres and other NO_x source areas; and (iii) that as much as 60–100% of the daily observed O_3 at any one place, e.g. coastal or mountain ridge sites, may result from fumigation from reservoir layers and/or advection within the recirculating air masses.

2.2 Ozone daily cycles, spatial and temporal variability: east coast of Spain as an example

Figure 10.3 shows the area under study in southern Europe and the monitoring network being considered. It also shows some of the observed O_3 diurnal cycles in July. Figure 10.3a illustrates the variation along a coastal valley with data from the MECAPIP project; Fig. 10.3b uses data from the network stations to illustrate the height dependence. The most notable features are the changes in the diurnal cycles from the sites at the coast to those at the mountain tops.

At the coast, the profiles look like square waves. A sharp rise in the morning is followed by nearly constant O_3 values during the afternoon, and a drop to low values at night. The mountain sites show a nearly flat profile with high concentrations day and night. Figure 10.4 shows the annual evolution of the O_3 diurnal cycles in this region, averaged for each month, for 1997. These

Fig. 10.2. Schematic diagram of the circulations in the coastal regions of the western Mediterranean in July. (a) Diurnal conditions. The letters a–d indicate successive stages in the development of the sea breeze and the formation of stratified reservoir layers aloft, and the numbers correspond to typical station sites. The lower graph represents the relationship between net ozone production and the amount of NO_x oxidized along the airmass trajectories, after Derwent and Davies (1994). (b) Nocturnal conditions. The ozone evolution along the path of the draining air is shown at the bottom. Stations located high above the coastal plains can remain within the reservoir layers during the night.

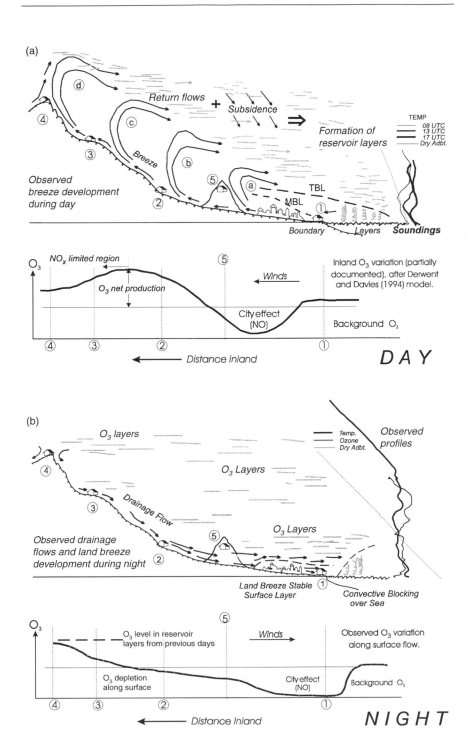

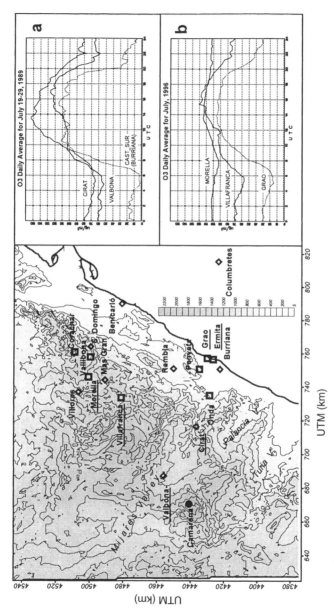

Fig. 10.3. Air quality monitoring network in the Castellón Province of Spain. Graphs (a) and (b) show average O₃ daily cycles for July at various altitudes and distances inland.

data document the existence of chronic-type episodes with persistent medium-to-high O_3 levels, as compared with shorter, peak-type, episodes in central and northern Europe. They also confirm the marked differences between the cycles observed in stations at the coast, mid-valley and mountain tops, and between some stations (Ermita/Grao and Penyeta) located just a few kilometres apart. All of these can lead to data interpretation problems for those responsible for the air quality networks and for the interpretation of ozone damage to vegetation.

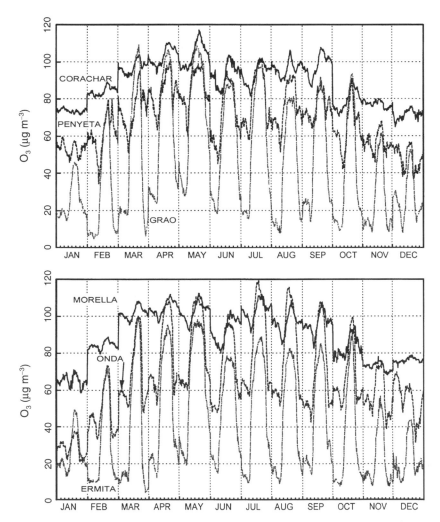

Fig. 10.4.   Monthly averaged O_3 diurnal cycles for 1997, at six stations on the Spanish Mediterranean coast located at various altitudes and distances from the coast (see Fig. 10.3).

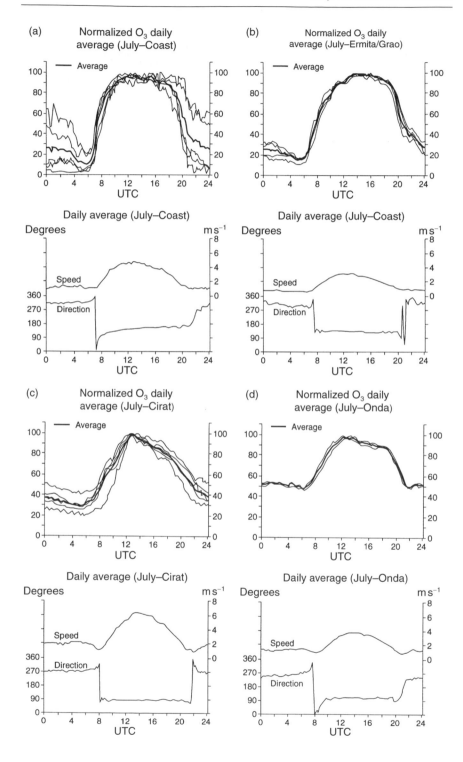

(a) Normalized O$_3$ daily average (July–Coast)

(b) Normalized O$_3$ daily average (July–Ermita/Grao)

Daily average (July–Coast)

Daily average (July–Coast)

(c) Normalized O$_3$ daily average (July–Cirat)

(d) Normalized O$_3$ daily average (July–Onda)

Daily average (July–Cirat)

Daily average (July–Onda)

This network was designed in 1993–1994 after the MECAPIP project documented the recirculations in this region, and was intended to: (i) compile a historical database of the processes along the same path used in the MECAPIP project; and (ii) support measurement campaigns for other EC projects in the western Mediterranean basin, e.g. Biogenic Emissions in the Mediterranean Area (BEMA) (Seufert, 1997), including measurements to validate coupled meteorological and photochemical models. The four stations at the northern edge of the domain began operation in 1995. The network includes fully automated stations (squares) which measure sulphur dioxide (SO_2), TSP (total suspended particulates), nitrous oxide (NO), nitrogen dioxide (NO_2), O_3 and meteorological parameters at standard 10 m height (temperature, wind speed, wind direction), and field sites (small diamonds). The latter have been instrumented and used during intensive measurement campaigns (1–2 weeks) in summer since 1989. The rest of the year the field sites are equipped with meteorological towers only.

2.3   A typology of observed O_3 cycles in July

To synthesize the daily variation and to emphasize the shape of the cycle instead of absolute values, the average cycles for each site and year have been normalized by dividing each value by the maximum and multiplying by 100. Thus, the vertical scale in Figs 10.5 and 10.6 is non-dimensional, and the solid line in these figures is the average of the normalized profiles. In relation to the stations in Fig. 10.3, and the schematics in Fig. 10.2, the typified cycles are as follows:

1.   Coastal stations (Burriana, Grao, Ermita, Benicarló, P. Sagunto). During the day they are immersed in the sea breeze which brings pre-mixed O_3 from the reservoir layers over the sea. In the evening and night the wind reverses, and these stations fall inside the drainage flows, which are enriched in NO emissions from the coastal strip and depleted in O_3. The observed O_3 cycles assume the form of a square wave, as shown in Fig. 10.5a and b. Data from intensive field campaigns (5–10 days) for the years 1989–1991 and 1994–1995 were used for Fig. 10.5a. Several sites were employed, all within a 4 km radius of Burriana (Fig. 10.4), and this explains some of the variations in nocturnal concentrations from year to year. For Fig. 10.5b data from the

---

Fig. 10.5. (opposite)   Normalized O_3 daily cycle (relative units), and averaged meteorological data for several sites in July. Coastal sites: Burriana (a) and Ermita-Grao (b). Valley floor sites: Cirat (c) and Onda (d). The data for Cirat are from the intensive field campaigns (5–10 days) in July, for the years 1989–1991 and 1994–1997.

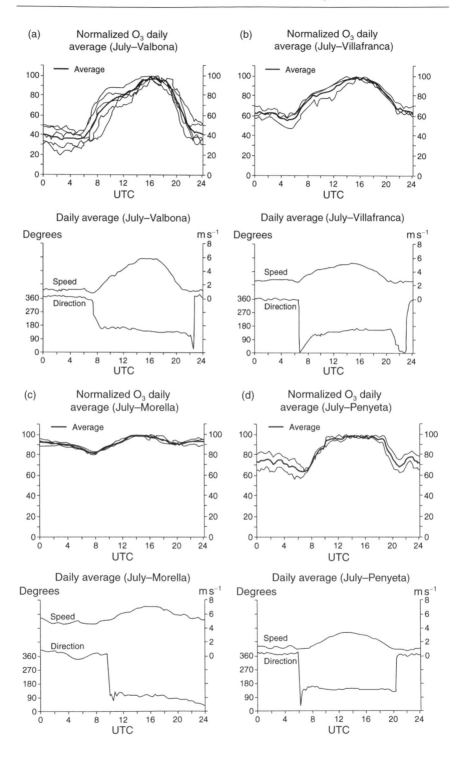

monitoring network were used. For the stations located on the coastal valley floor at varying distances from the coast, two locations can be distinguished.

2.   Valley floor within the O_3-producing region (Onda, Cirat, S. Domingo). Their cycle shows a morning rise, due to fumigation of old O_3 from the reservoir layers, which merges almost directly with another rise and the diurnal maximum from new O_3 produced photochemically within and advected by the developing sea breeze. During the night they sample drainage air partially depleted of O_3 (Fig. 10.5c, d).

3.   Upper-valley floor in the NO_x-limited region (Valbona, Villafranca). During the day they are within the path of the pollutants coming from the coast, and during the night they sample drainage air partially depleted of O_3. Their cycle shows a morning rise from fumigation of O_3 from the reservoir layers. This is followed by a clearly marked shoulder and a second rise with a maximum when the sea-breeze front brings fresh O_3 from the coast (Fig. 10.6a, b).

4.   Mountain ridges (Corachar, Morella, Vallibona, Camarena). During the day they form part of the orographic 'chimneys' that inject pollutants into the reservoir layers aloft. During the night they remain inside the upper reservoir layers. Their diurnal cycle is dampened by high O_3 concentrations (Fig. 10.6c). During the night these sites are also sensitive to interregional transport of O_3 layers, including those injected elsewhere. Their diurnal minima coincide with the maximum rise at the stations in the valleys and coastal plains.

5.   High coastal sites (Penyeta). Stations located high above the coastal plain, or on up-valley slopes, and near the coast. The observed O_3 cycles vary as a function of their altitude over the coastal plain and altitude determines whether they remain inside the lower reservoir layers, or are eventually engulfed by the stable surface layer during the night. During the day they may show the flat top characteristic of the coastal stations (Fig. 10.6d).

The shape of the observed cycles remains similar at all sites, in spite of the inter-annual variations in the absolute values of O_3 concentrations. This also emphasizes that the meso-meteorological circulations, which have always been an integral part of the Mediterranean climate system, are maintained; i.e. they are repeated for each site every year, even though the chemical production, dependent on emissions, may be on the increase and varies from year to year. It also emphasizes that meso-meteorological processes dominate the (repetitive) shape of the observed O_3 cycles in summer.

Fig. 10.6. (opposite)   Normalized O_3 daily cycle (relative units), and averaged meteorological data for several sites in July. Upper valley floor sites: Valbona (a) and Villafranca (b). Mountain ridge site: Morella (c) and a high coastal site: Penyeta (d). The graphs combine data from the 1989–1991 intensive field campaigns and continuous data for July 1994–1995 (Valbona), and from the automatic station in operation since 1995.

2.4   Ozone daily cycles in the western Mediterranean: general statements

In the western Mediterranean basin the observed O_3 cycles depend strongly on the topographic location of the observing station and its relationship to the reservoir layers, the atmospheric circulations involved, and the chemical processes along each path. Thus, each O_3 monitoring station shows a part of the whole, and could even be considered to represent a specific area, providing that the relevant processes are understood for each site, and that the site itself has been adequately selected. However, no single station can be considered representative of regional processes, and much less of the whole situation. The modelling of atmospheric dispersion for any Mediterranean coastal site can lead to questionable results if one does not take into account that the evolution of the boundary layer can be dominated by meso-scale processes acting at distances of from 100 to 400 km from the coast (Millan *et al.*, 1999).

2.5   Modelling the ozone in the Mediterranean

Few, if any, of the atmospheric processes and their effects that are now known to exist were known in these regions before 1986, or even 1992, despite large international measurement programmes (e.g. EMEP) and other extensive modelling efforts. Thus, the use of these same models, with typical 150 km grids (or even with 50 km grids) for regulatory purposes, policy decisions or economic evaluations, should be seriously and consistently questioned until they are made to reflect the true workings of the system. Modelling sensitivity studies indicate that for this area of complex coastal terrain grid resolutions of 2 km, or less, are required.

Most monitoring stations in southern European cities were based on health criteria related to SO_2 and particulates, and O_3 monitoring networks which include rural stations are therefore still sparse in the region. In the meantime, modelling, if used with the proper spatial resolution, could be used to describe the physicochemical processes in this region (Bastrup-Birk *et al.*, 1997; Thunis and Cuvelier, 1997).

3   Ozone and forest ecosystems

3.1   Air pollution problems today, with emphasis on the Mediterranean region

Outside North America and central and northern Europe, research has concentrated primarily on pollutants such as fluoride, SO_2, nitrogen deposition and O_3 because they present the most immediate problems for forest health.

However, in Europe, considerable changes are occurring in the pollution climate as a result of the steps being taken to reduce the emissions of certain pollutants (Chapter 12, this volume). The importance of sulphur dioxide as a pollutant is declining over much of the continent, including the Mediterranean area (Avila, 1996), and its effects on plants are being examined in fewer and fewer experiments. In contrast, nitrogen emissions have increased (Fowler *et al.*, 1998) and remain of considerable importance, firstly because of potential ecosystem eutrophication, and secondly because the oxides of nitrogen are an important precursor for ozone. Today, fluoride is less important than before, but it still remains a problem in some areas (Eleftheriou and Tsekos, 1991; Brigati *et al.*, 1995), as does the deposition of heavy metals (cf. the chapters in this volume on the Kola Peninsula).

Of the various pollutants present in Europe, ozone has generally been considered one of the most important for several years, particularly in the Mediterranean area. Thus, one of today's major environmental concerns is to understand the atmospheric processes that control tropospheric ozone and OH-radical budgets. Trace gas exchanges with terrestrial ecosystems and their potential effects on ecosystems are not well understood. For instance, more information on biogenic emissions (e.g. terpenes and isoprene) from forests and their potential role in ozone formation is needed, although some large EC projects (e.g. BEMA) have improved our understanding of biogenic emissions in the Mediterranean area and their potential role in tropospheric O_3 formation (Seufert, 1997). Feedback interactions between ozone and trace gas emissions from plants seem to be an important unexplored issue, as results suggest that biogenic trace gases such as ethylene and some sesquiterpenes increase with any increase in O_3 concentrations (Mehlhorn and Wellburn, 1987; Schuh *et al.*, 1996). As a result of photochemical processes, the deposition of nitrogen via the atmosphere seems to be increasing (Loye-Pilot *et al.*, 1989; Fowler *et al.*, 1998). Thus modifications of the N supply might directly influence carbon cycling and may lead to limitations in the supply of other major or minor nutrients, especially phosphorus.

3.2   Critical levels for ozone

The concept and its application to the Mediterranean region

In principle, any useful exposure index must be based on the concept of effective dose, i.e. it must capture the characteristics of exposure which most directly relate to the amount of ozone that is absorbed by vegetation (by multiplying the concentration near the leaf surface by the leaf conductance for ozone). Due to the lack of leaf conductance data, the use of radiation as a surrogate for leaf conductance has been suggested in agricultural crops, and the simplest approach is to use ozone concentrations measured during daylight hours (i.e. > 50 W m^{-2} global radiation) to characterize exposure. For species with substantial leaf conductance at night, however, no such discrimination should be made, particularly in view of the results of Matyssek *et al.* (1995). Other factors, such as air humidity, soil water availability and temperature, are also known to influence leaf conductance, but to date these factors have not been used to characterize ozone uptake or dose in long-term experiments. However, such factors are very important in the modulation of the photosynthetic response of Mediterranean vegetation, as has been shown for some Mediterranean trees (Naveh *et al.*, 1980; Inclán and Gimeno, 1998).

Long-term exposure to ozone can lead to growth and yield reductions, so the most suitable exposure indices related to long-term effects are cumulative (Lee *et al.*, 1988; Lefohn, 1992; Finnan *et al.*, 1997). Experimental exposure–response studies with ozone suggest that intermittent exposure to higher concentrations is the most important in causing long-term effects. It is therefore essential to consider both the duration and the extent of exceedance of any guideline value, not just the total exceedance. This is an important issue if the different climatic conditions and resulting O_3 patterns in the Mediterranean are to be taken into account, as the predominance of relatively high chronic concentrations has a significant impact on any exposure indices (Millán *et al.*, 1997, 1999). However, the current O_3 Directive of the EC (Directive 92/72/CEE) only includes a given 24 h mean (65 μg m^{-3}) as a threshold to protect plants.

In recent years more data have become available from Europe, highlighting that the AOT40 index (accumulated hours above 40 ppb) appears to be suitable to relate exposure to plant response, at least for crops such as wheat (Fuhrer *et al.*, 1997). Furthermore, it seems to work quite well with experimental data. As a result, the AOT40 index with various thresholds has been recommended by the United Nations Economic Council for Europe for the protection of different types of vegetation (Fuhrer and Achermann, 1994). Although the use of critical levels for ozone is now primarily associated with the AOT40 index, following agreement at the 1993 Bern Workshop (Fuhrer and Achermann, 1994), two distinct types of values can be identified. Level I involves a simple risk assessment without taking into account any modifying factors. This is the use of the AOT40 index without any qualification, as for example in maps or diagrams of AOT40–10,000 ppbh exceedances for forest areas.

Level II involves the incorporation of a range of modifying factors (e.g. air humidity or phenological stage) and is much more difficult to establish. However, only Level II provides meaningful information for the assessment of the potential impacts of ozone. Critical research issues surrounding the definition of this level remain, including the measurement and characterization of the atmospheric exposure at the plant surface, the stomatal uptake of pollutant, the appropriate threshold value, the length of the AOT calculation period, the incorporation of night-time values in addition to day-time values, the choice of which months to include, possible phenological effects, and the differential effects according to phenotype.

A more recent workshop, 'Critical levels for ozone in Europe: testing and finalising the concepts' held in Kuopio, Finland (Kärenlampi and Skärby, 1996), identified deciduous trees as the most sensitive receptors. A value of 40 ppb was chosen as the cut-off level for calculating a biologically meaningful index of ozone exposure. The response parameter chosen was the percentage reduction in annual biomass increment. A critical level of 10,000 ppbh over 40 ppb was accepted. This threshold is exceeded over much of Europe. In the Mediterranean countries and surrounding mountainous areas, the available ozone data indicate that this index is exceeded repeatedly, and injury and damage should be expected throughout the region. However, such effects have yet to be described in this area and we do not even know if Mediterranean species are as sensitive as anticipated.

The new EC Directive 96/62/EC on ambient air quality assessment and management committed the 15 member states of the EC to review all air quality legislation standards, and the future Daughter Directive of Ozone for vegetation protection will be established at the end of 1999. For ozone, this could be the opportunity to introduce a parameter which closely relates exposure to plant response. However, the recommendations to have the growing season as a fixed window (April to September) for forests may be inappropriate for Mediterranean countries, where phenology may be different and several other factors are very important (e.g. moisture and radiation).

Evidence of critical level exceedances in Europe, with emphasis on the Mediterranean region

The evidence that we have to date strongly suggests that ozone occurs at sufficient concentrations to cause injury to sensitive species, including trees. Maps of the cumulative exposure index AOT40 for Switzerland indicate that exceedances occur over very large areas, particularly in the important (for forests) zone at 1000–1500 m a.s.l. The threshold value of AOT40–10,000 (ppbh) for the April–September growing season is exceeded every year at most Swiss monitoring stations, with the highest exceedances being reported from a subalpine mountain site on Rigi (NABEL, 1995). In Mediterranean countries exceedances of the AOT40 in forested areas have also been recorded; most of the rural stations on the eastern coast of Spain up to 100 km inland exceeded this threshold, as did most of the mountain monitoring stations above 1000 m (Sanz and Calatayud, 1999; Sanz et al., 1998b; Fig. 10.7). Similar data are available from Switzerland (Innes et al., 1996) and Italy (Bussotti and Ferretti, 1998).

Although there are some discrepancies between modelled and measured values in Mediterranean countries such as Spain (EMEP, 1995, 1998), probably due to the lack of data to adjust the models properly in these countries, the measured and modelled exceedances of AOT40 for crops and semi-natural vegetation show that the threshold of 3000 ppbh is exceeded in most of the EC member states and there are also exceedances for AOT40–10,000 ppbh for forests (EMEP, 1995). Exceedances by a factor of five above the threshold are experienced in southern Europe. When the 5-year averaged modelled excess O_3 over the threshold for forests (10,000 ppbh) is plotted, it reveals a north to

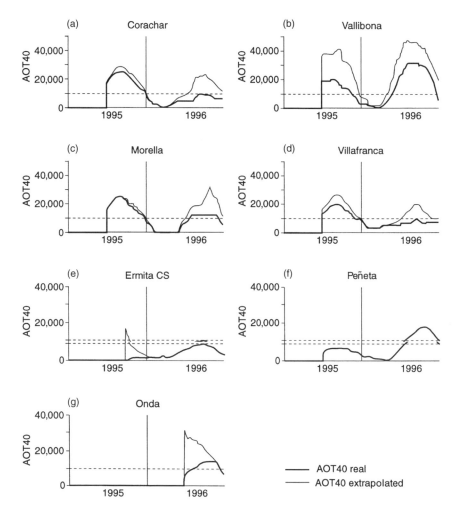

Fig. 10.7. Running AOT40 for 6 months as recommended in Kärenlampi and Skärby (1996) for 1995 and 1996 from the Air Quality Network stations near sampling sites. (a), (b), (c), (d), at the top of the mountains near the Bergantes and Cérvol valleys. (e), (f), (g) in the Mijares valley. AOT40 of 10,000 ppbh for 6 months is the threshold recommended for forest protection. The thick line represents calculations with real data. The thin line represents data adopting a linear extrapolation when hourly means are missing.

south gradient. Other modelling efforts show even higher exceedances in the Mediterranean region (Bastrup-Birk *et al.*, 1997).

This has led to the expectation that widespread damage should be occurring in forests as a result of O_3. Large-scale inventories of crown condition have been undertaken since the mid-1980s, and these have revealed a progressive increase in crown transparency. This increase is present in southern Europe, but it may mask a variety of regional and species-specific trends, some of which are known to have nothing to do with air pollution. A good review of the available information in southern Europe is provided by Bussotti and Ferretti (1998). Attempts to undertake international analyses of the data have been complicated by the presence of major differences in the assessments made by observers in different countries, although this problem has to some extent been successfully addressed in southern Europe (e.g. Ferretti *et al.*, 1997). The greatest problem, however, lies in the emphasis placed on a non-specific symptom of tree condition, namely crown transparency (Ferretti, 1997). Thus, it will be very useful to add some specific symptoms for pollutants such as ozone to the assessments. First, however, they must be adequately characterized.

3.3   Ozone damage, what and how?

It is only in the last decade that the potential impacts of ozone have become of concern in southern Europe. There is some evidence that the ambient ozone concentrations found in Europe can cause a range of effects, including visible leaf injury, growth and yield reductions, and altered sensitivity to biotic and abiotic stresses. Recent research has advanced our understanding of the underlying mechanisms of ozone effects on agricultural crops, and to a lesser extent on trees and native plant species. Long-term effects on trees may impair the function of forest ecosystems, i.e. their role with respect to water and energy balances, to the protection of soil against erosion, vegetation cover in dry areas, etc. Some of the most important impacts on plant communities may be through shifts in species composition, loss of biodiversity and changes in genetic composition, particularly as southern Europe has large numbers of endemic plant species.

Some of the mechanisms leading to the development of ozone injury are quite well known, although several uncertainties remain. Plant cuticles can be damaged by ozone, but relatively little O_3 penetrates the leaf via the cuticle. Instead, most O_3 enters plants through the stomatal pores on the leaf surface where it then passes into the substomatal intercellular spaces (Guderian *et al.*, 1985). Consequently, any factors affecting stomatal aperture will influence the plant tissue response to O_3. Moreover, stomatal response to environmental stresses may reduce O_3 uptake under certain climatic conditions (such as the dry summers typical of southern Europe). Contrary to earlier ideas, it is now believed that O_3 does not have a direct effect on stomatal conductance

(Weber *et al.*, 1994) or at least this is not its main effect, although high peak values may have a direct effect (see, for example, the work of Minnocci *et al.*, 1995).

The most sensitive tissues within leaves are the photosynthetically active cells containing chloroplasts, and the first site of injury is usually the cell walls and plasma membranes (Guderian *et al.*, 1985). Palisade cells tend to be more sensitive than mesophyll cells (Tiedemann and Herrmann, 1992). In species showing visual symptoms, it is the palisade cells that are usually affected first. The most frequent symptoms of ozone injury are chlorosis and necrotic stipple, which develop if repair mechanisms cannot keep up with the cellular injury. The repair mechanisms may be associated with the synthesis of ethylene within the cells (Langebartels *et al.*, 1991) and the activation of extracellular peroxidases (Heath, 1994). Repair mechanisms may prevent foliar injury from arising, but the physiological costs of these repairs may induce damage by diverting resources away from growth (Reich, 1983; Wolfenden and Mansfield, 1991). Physiological changes such as reduced photosynthesis may also occur, which can lead to a reduction in yield or an alteration in the allocation of biomass between the leaves, roots and seeds without the appearance of visual leaf injury. These changes are less obvious and more difficult to quantify in relatively unproductive and slow-growing forests, as is the case in most of the Mediterranean region.

Experimental vs. field data, seedlings vs. adult trees, one species vs. the ecosystem

The extrapolation of seedling data to larger trees is complicated by differences in the percentage of photosynthetically active tissue and associated changes in gas exchange properties, changes in carbon allocation and use patterns, and altered water and nutrient regimes (Kramer and Kozlowski, 1979; Edwards *et al.*, 1994). There are relatively few extrapolations from individual plants to the stand, and community-level assessments are also at an early stage of development (Patterson and Rundel, 1995). Consequently, although there have been several significant developments related to scaling (e.g. Ehleringer and Field, 1993; Schulze *et al.*, 1994), effects on successional processes, growth and productivity, competition and biodiversity, as a few examples, remain poorly understood, and are particularly complex in Mediterranean ecosystems.

The data used to derive critical levels are almost entirely drawn from experiments in open-top chambers in central and northern Europe, using plants that are adequately supplied with water and nutrients. For trees, even experimental data are scarce and many species have not been investigated in Europe (especially in the Mediterranean region). Considerable difficulties also exist in scaling-up from seedlings to mature trees. Consequently, the available exposure–response data from studies with seedlings may not reflect the response to O_3 of older, mature trees or forest stands, and the effects of long-term exposures on trees with a long life cycle are unknown (Matyssek *et al.*, 1995).

Visual symptoms as a tool in field surveys

Until now, most experiments have concentrated on the explanation of the mechanisms leading to the injury observed in experimental studies, rather than trying to explain symptoms observed in the field. With the main exception of documented foliar symptoms attributable to specific air pollutants (e.g. adaxial leaf stipple and tropospheric ozone), the cause–effect determinations normally applied in biological research (e.g. Koch, 1876) have often been laid aside. In contrast, numerous early studies of foliar injury caused by ozone in California (e.g. Miller and Parmeter, 1967; Miller and Millecan, 1971; Miller, 1973) and several more recent studies (e.g. Prinz *et al.*, 1982; Liebold, 1988; Davis and Skelly, 1992; Simini *et al.*, 1992) have linked laboratory exposure-based work with forest observations under conditions of ambient exposures. Such studies are clearly the way forward.

It is not possible to say whether O_3 has had an influence on the two main indicators of crown condition, discoloration and transparency. For some species, such as black cherry (*Prunus serotina* Ehrh.), a non-native European species known to be sensitive to ozone and showing clear early senescence of leaves, there appears to be a relationship. Similar symptoms have also been observed on native species in Switzerland and Spain (e.g. common ash, *Fraxinus excelsior* L. and European beech, *Fagus sylvatica* L.) and on non-native species (e.g. false acacia, *Robinia pseudoacacia* L. and tree of heaven, *Ailanthus altissima* (Mill.) Swingle), and chlorotic mottle has been recorded on Aleppo pine (*Pinus halepensis* Mill.) at a number of locations in the Mediterranean. The evidence that we have to date strongly suggests that ozone occurs at sufficient concentrations to cause visible foliar injury to sensitive trees. Even if visual injury does not include all possible forms of injury to trees and natural vegetation, it may be a useful tool to detect ozone injury in potentially sensitive species in Europe during extensive field surveys.

3.4   Examples of ozone effects in the Mediterranean region

Aleppo pine in the east of Spain

In recent years, there has been an increasing concern about the condition of Aleppo pine in the Mediterranean region (Gimeno *et al.*, 1992; Velissariou *et al.*, 1992; Garrec, 1994; Davison *et al.*, 1995; Sanz and Millán, 1998). On the east coast of Spain, this species shows a high level of crown transparency which is associated with low needle retention. Gimeno and Elvira (1996) suggested that an AOT40 of 45,000 ppbh over 6 months is necessary to reduce the photosynthesis rate by 20% in 4-year-old Aleppo pine seedlings, although with no appreciable growth reductions. In the field, however, symptoms appeared after 10,000 ppbh exposure (Sanz et al., 1998a).

Ozone injury on Aleppo pine is quite widespread in eastern Spain. Epistomatal wax fusion and chlorotic mottle in Aleppo pine during 1994 were well correlated with the penetration of sea breezes in the coastal valleys of Castellón (Spain) (Sanz and Millán, 1998). More recent data revealed a gradient of O_3 injury (chlorotic mottle) and wax fusion along other coastal valleys (Sanz et al., 1998a; Sanz and Calatayud, in press). Higher percentages of chlorotic mottle and wax fusion occur in the needles of Aleppo pine from the lower parts of the valleys, where sea-breezes occur more frequently throughout the year. These sea-breezes lead to high ozone levels inland from the coast. Symptoms have been recorded in several valleys on the east coast of Spain in 1995, 1996 and 1997: Mijares valley (Castellón-Teruel), Palancia Valley (Valencia), Serpis Valley (Alicante), including protected areas such as natural parks (Sanz and Millán, 1998; Sanz and Calatayud, in press; Sanz et al., unpublished data). Symptoms have also been recorded by Davison et al. (1995) along the coast between Tarragona and Castellón. Results from an intensive survey along the Mijares valley, looking at phenology, O_3-like symptoms, pests and diseases, running since 1996, revealed that the older the needles were, the more chlorotic mottle they showed. Chlorotic mottle in the needles appeared for the first time in early summer (in 1-year-old needles); once this symptom has appeared its apparent intensity in a given whorl of needles can change throughout the year (Hernández et al., 1998). These results are consistent with the observed dynamics of the polluted air masses in this Mediterranean region, including transport via the Ebro Valley.

There is a strong interannual fluctuation in ozone concentrations, with consequent variability in O_3 injury from year to year (Sanz et al., 1998b; Sanz and Calatayud, in press; Millán et al., in press). In the same area, factors such as drought and disease also occur, in addition to the pigment changes (Davison et al., 1995), producing more complex dynamics. 'Carry-over' effects were seen in the response of Aleppo pine to O_3 during controlled experiments, which may enhance the sensitivity of the needles during subsequent exposures to O_3 (Elvira et al., 1998). This type of chlorotic mottle is similar to that observed in ponderosa pine (Pinus ponderosa Dougl.) in California (Miller et al., 1963, 1969; Jacobson and Hill, 1970) and it has therefore been related to the effects of photooxidants (Richards et al., 1968; Naveh et al., 1980; Kärenlampi, 1987; Gimeno et al., 1992; Velissariou et al., 1992; Davison et al., 1995).

Ozone-like symptoms in other species in Spain and Switzerland

The results of surveys of native species in areas of eastern Spain and across broad areas of Switzerland strongly suggest that numerous native plant species are exhibiting ozone sensitivity in the form of typical foliar symptoms (Skelly *et al.*, 1999). In 1996, a survey looking for symptoms typical of ozone injury was conducted in some of the valleys in Spain listed above. In all, 49 species of trees, shrubs and herbs had ozone-like symptoms. Some of the species are already known to be sensitive to ozone, such as tree of heaven and false acacia. Also almond trees (*Prunus dulcis* D.A. Webb) growing in plantations showed symptoms. Numerous differing species within genera common to both Switzerland and Spain have been noted as symptomatic, with adaxial foliar stippling being the primary symptom. The most prominent genera include *Viburnum, Cornus, Rumex, Fraxinus, Epilobium* and *Prunus* (see Table 10.1).

Table 10.1. Native plant species of southern Spain observed with ozone-like symptoms during surveys of natural habitats and forest nursery conditions in the late summer seasons (1995–1998) (from Skelly *et al.*, 1999).

Shrub species	Tree species	Forb species
Anthyllis cytisoides L.*	*Ailanthus altissima* (Mill.) Swingle, i, **	*Abutilon theophrasti* Medik., i
Arbutus unedo L.*	*Cornus sanguinea* L.*	*Agrimonia eupatoria* L.
Cistus salvifolius L.*	*Corylus avellana* L.	*Calystegia sepium* (L.) R. Br.
Lonicera etrusca G. Sant.	*Cytisus patens* L.*	*Campanula* sp.
Lonicera implexa Aiton*	*Fraxinus ornus* L.	*Chamerion angustifolium* (L.) Holub.**
Myrtus communis L.*	*Juglans regia* L.	*Chenopodium album* L.
Pistacia lentiscus L.*	*Morus nigra* L.	*Colutea arborescens* L.
Pistacia terebinthus L.*	*Populus alba* L.	*Dittrichia viscosa* (L.) Greuter
Prunus spinosa L.	*Populus nigra* L.	*Epilobium collinum* Gmelin*
Ricinus communis L.*, **	*Prunus dulcis* D.A. Webb**	*Ipomea sagittata*
Rhamnus alaternus L.	*Robinia pseudoacacia* L.	*Langestroemia indica*, i
Rosa cf. *canina* L.		*Oenothera rosea* L.'Hér. ex Aiton**
Rubia peregrina L.*		*Parthenocissus quinquefolia* (L.) Planch.
Rubus ulmifolius Schott		*Plantago lanceolata* L.
Sambucus nigra L.		*Rumex pulcher* L.**
Viburnum tinus L.		*Verbascum sinuatum* L.**
		Vinca difformis Pourr.*

*Confirmed via OTC studies in Valencia, Spain 1997, 1998.
**Confirmed via CSTR chamber investigations, Penn State University.
i, introduced.

Records of ozone visible injury in other native species in Italy and Greece

Several surveys have been performed in the Mediterranean region to assess the extent of visible injury induced by ozone, mostly concentrated on crops in Tuscany, Italy (Lorenzini *et al.*, 1984), eastern Spain and Greece (Velissariou *et al.*, 1996). Less information is available about native tree species, but ozone visible injury has been reported on hybrid poplar in Italy (Schenone, 1993) and on Austrian pine (*Pinus nigra* Arnold) and Grecian fir (*Abies cephalonica* Loud.) in Greece (Velissariou *et al.*, 1996).

4 Conclusions

The evidence that we have to date strongly suggests that O_3 occurs at sufficient concentrations to cause injury to sensitive species, including trees, in the Mediterranean area. This is because under strong summer insolation, the coastal recirculations become 'large natural photochemical reactors' where most of the NO_x emissions and other precursors are transformed into oxidants, acidic compounds, aerosols and O_3.

Foliar injury caused by ambient O_3 exposures has been clearly identified on native herbs, shrubs and tree species within several areas in southern Europe. Combined studies in southern Switzerland and along the east coast of Spain have involved initial field observations followed by controlled exposures and have confirmed that most of the observed visual injuries were produced by O_3. There are also similar records from Italy and Greece, as well as from Spain, from other studies. In southern Europe, crops appear to show visual symptoms more often, and over larger areas, than native vegetation and trees. This may indicate that ozone uptake and response is modulated by environmental factors such as drought. Recent studies suggest a complex interaction between high radiation, drought and ozone response in Mediterranean tree species such as Aleppo pine.

Critical levels for ozone visual injury are exceeded in southern Europe and natural vegetation (e.g. forests) is therefore at risk. However, there are insufficient data to assess the extent of this risk, and more research is needed to better understand and characterize the response of Mediterranean plants to ozone, and to understand the ways in which environmental factors modify the responses of plants exposed to ozone.

Acknowledgements

The work presented here was financed by the European Commission, under different projects dealing with atmospheric dynamics and atmospheric chemistry. The Generalidad Valenciana, CICYT and BANCAIXA also supported part of the projects. We thank Dr J. Innes and Dr J. Skelly for their help and comments, as well as Mr G. Sánchez and Dr R. Montoya from the DGCONA for their support.

References

Avila, A. (1996) Time trends in the precipitation chemistry at a mountain site in Northeastern Spain for the period 1983–1994. *Atmospheric Environment* 30, 1363–1373.

Bastrup-Birk, A., Brandt, J., Zlatev, Z. and Uria, I. (1997) Studying cumulative ozone exposures in Europe during a 7-year period. *Journal of Geophysical Research* 102, 23917–23935.

Brigati, S., Maccaferri, M. and Dondi, M. (1995) Phytotoxic symptoms in pine due to fluoride emissions. In: Lorenzini, G. and Soldatini, G.F. (eds) *Responses of Plants to Air Pollution*. Agricoltura Mediterranea, special volume, pp. 221–223.

Bussotti, F. and Ferretti, M. (1998) Air pollution, forest condition and forest decline in Southern Europe: an overview. *Environmental Pollution* 101, 49–65.

Butkovic, V., Cvitas, T. and Klasing, L. (1990) Photochemical ozone in the Mediterranean. The *Science of the Total Environment* 99, 145–151.

Ciccioli, P., Brancaleoni, E., Di Paolo, C., Brachetti, A. and Cecinato, A. (1987) Daily trends of photochemical oxidants and their precursors in a suburban forested area. A useful approach for evaluating the relative contributions of natural and anthropogenic hydrocarbons to the photochemical smog formation in rural areas of Italy. In: Angeletti, G. and Restelli, G. (eds) *Physico-Chemical Behavior of Atmospheric Pollutants. Proceedings of the 4th European Symposium*. D. Reidel Publishing, Dordrecht, pp. 551–559.

Davis, D.D. and Skelly, J.M. (1992) Growth response of four species of eastern hardwood tree seedlings exposed to ozone, acidic precipitation, and sulfur dioxide. *Journal of the Air Pollution Control Association* 42, 309–311.

Davison, A.W., Velissariou, D., Barnes, J.D., Gimeno, B. and Inclan, R. (1995) The use of Aleppo pine, *Pinus halepensis* Mill as a bioindicator of ozone stress in Greece and Spain. In: Munawar, M., Hänninen, O., Roy, S., Munawar, N., Kärenlampi, L. and Brown D. (eds) *Bioindicators of Environmental Health*. Ecovision World Monograph Series, SPB Academic Publishing, Amsterdam, pp. 63–72.

Derwent, R.G. and Davies, T.J. (1994) Modelling the impact of NO_x or hydrocarbon control on photochemical ozone in Europe. *Atmospheric Environment* 28, 2039–2052.

Edwards, G.S., Wullschleger, S.D. and Kelly, J.M. (1994) Growth and physiology on northern red oak: Preliminary comparisons of mature tree and seedling responses to ozone. *Environmental Pollution* 83, 215–221.

Ehleringer, J. and Field, C.B. (1993) *Scaling Physiological Processes: Leaf to Globe*. Academic Press, New York.

Eleftheriou, E.P. and Tsekos, I. (1991) Fluoride effects on leaf cell ultrastructure of olive trees growing in the vicinity of the aluminum factory of Greece. *Trees* 5, 83–88.

Elvira, S., Alonso, R., Castillo, F. and Gimeno, B.S. (1998) On the response of pigments and antioxidants of *Pinus halepensis* seedlings to Mediterranean climatic factors and long-term ozone exposure. *New Phytologist* 138, 419–432.

EMEP (1995) *Transboundary Photooxidant Air Pollution in Europe: Calculations of Tropospheric Ozone and Comparison with Observations*. EMEP/MSC-W Report 2/95, NMI, Blindern, Norway.

EMEP (1998) *Transboundary Photooxidant Air Pollution in Europe: Calculations of Tropospheric Ozone and Comparison with Observations*. EMEP/MSC-W Report 2/98, NMI, Blindern, Norway.

Ferretti, M. (1997) Forest health assessment and monitoring – issues for consideration. *Environmental Monitoring and Assessment* 48, 45–72.

Ferretti, M., Cozzi, A. and Cenni, E. (1997) Importanza delle procedure di Quality Assurance nel monitoraggio degli ecosistemi. Le foreste come esempio. In: Anelli, A., Ferrari, I., Rossetti, G. and Vezzosi, M. (eds) *Ecologia – Atti dell'VIII Congresso Nazionale della Società Italiana di Ecologia, Parma, 10–12 Settembre 1997*. Società Italiana di Ecologia Atti 18, 323–326.

Finnan, J.M., Burke, J.I. and Jones, M.B. (1997) An evaluation of indices that describe the impact of ozone on the yield of spring wheat (*Trictium aestivum* L.). *Atmospheric Environment* PII, 2685–2693.

Fortezza, F., Strocchi, V., Giovanelli, G., Bonasoni, P. and Georgiadis, T. (1993) Transport of photochemical oxidants along the northwestern Adriatic coast. *Atmospheric Environment* 27A, 2393–2402.

Fowler, D., Flechard, C., Skiba, U., Coyle M. and Cape, N. (1998) The atmospheric budget of oxidized nitrogen and its role in ozone formation and deposition. *New Phytologist* 139, 11–23.

Fuhrer, J. and Achermann, B. (eds) (1994) *Critical Levels for Ozone – a UN-ECE Workshop Report*, Schriftenreihe der FAC Liebefeld 16. Swiss Federal Research Station for Agricultural Chemistry and Environmental Hygiene, Liebefeld-Bern.

Fuhrer, J., Skärby, L. and Ashmore, M.R. (1997) Critical levels for ozone effects on vegetation in Europe. *Environmental Pollution* 97, 91–106.

Garrec, J-P. (1994) Les dépérissements litoraux d'arbres forestiers. *Revue Forestière Française* 46, 454–457.

Georgiadis, T., Giovanelli, G. and Fortezza, F. (1994) Vertical layering of photochemical ozone during land-sea breeze transport. *Il Nuovo Cimento* 17, 371–375.

Gimeno, B.S. and Elvira, S. (1996) A contribution to the set-up of ozone critical levels for forest trees in Mediterranean areas. Results from the exposure of Aleppo pine (*Pinus halepensis* Mill.) seedlings in open top chambers. In: Kärenlampi, L. and Skärby, L. (eds) *Critical Levels for Ozone in Europe: Testing and Finalizing the Concepts*. *UN-ECE Convention on Long-Range Transboundary Air Pollution Workshop Report*. Department of Ecology and Environmental Science, University of Kuopio, Kuopio, pp. 169–182.

Gimeno, B.S., Velissariou, D., Barnes, J.D., Inclán, R., Peña, J.M. and Davison, A. (1992) Daños visibles por ozono en aciculas de *Pinus halepensis* Mill. en Grecia y España. *Ecologia* 6, 131–134.

Guderian, R., Tingey, D.T. and Rabe, R. (1985) Effects of photochemical oxidants on plants. In: Guderian, R. (ed.) *Air Pollution by Photochemical Oxidants*. Springer Verlag, New York, pp. 129–333.

Heath, R.L. (1994) Possible mechanisms for the inhibition of photosynthesis by ozone. *Photosynthesis Research* 39, 439–451.

Hernández, R., Pérez, V., Sánchez, G., Calatayud, V., Sanz, M.J. and Calvo, E. (1998) Needle injury in Aleppo pine: a long term study in the east coast of Spain.

Presented at the IUFRO Meeting, Forest Growth Responses to the Pollution Climate of the 21st Century, Edinburgh, Scotland, 21–23 September 1998.

Inclán, R. and Gimeno, B.S. (1998) Ozone and drought stress interactive effects on the physiology of Aleppo pine (*Pinus halepensis* Mill.). *Chemosphere* 36, 685–690.

Innes, J.L., Skelly, J.M., Landolt, W., Hug, C., Snyder, K.R. and Savage, J.E. (1996) Development of visible injury on the leaves of *Prunus serotina* in Ticino, southern Switzerland, as a result of ozone exposure. Preliminary results. In: Knoflacher, M., Schneider, J. and Soja, G. (eds) *Exceedance of Critical Loads and Levels. Report of a Workshop Held in Vienna, Austria Under the Convention on Long Range Transboundary Air Pollution, 22–24 November 1995*. Conference Papers No. 15, Bundesministerium für Umwelt, Jugend und Familie, Vienna, pp. 146–154.

Jacobson, J.S. and Hill, A.C. (1970) *Recognition of Air Pollution Injury to Vegetation: a Pictorial Atlas*. Air Pollution Control Association, Pittsburgh.

Kärenlampi, L. (1987) Visible symptoms and mesophyll cell structural responses to air pollution in the lowland pines (*Pinus radiata* and *P. halepensis*) in Southern California. *Savonia* 9, 1–12.

Kärenlampi, L. and Skärby, L. (1996) *Critical Levels for Ozone in Europe: Testing and Finalizing the Concepts. UN-ECE Convention on Long-Range Transboundary Air Pollution Workshop in Kuopio*. Department of Ecology and Environmental Science, University of Kuopio, Kuopio.

Koch, R. (1876) Untersuchung über Bacterien. V. Die etiologie der Milzbrandkrankheit, begründet auf die Entwicklungsgeschichte des Bacillus Anthracis. *Beiträge zur Biologie der Pflanzen* 2, 277–310.

Kramer, P.J. and Kozlowski, T.T. (1979) *Physiology of Woody Plants*. Academic Press, New York.

Langebartels, C., Kerner, K., Leonardi, S., Schraudner, M., Trost, M., Heller, W. and Sandermann, H. (1991) Biochemical plant responses to ozone. I. Differential induction of polyamine and ethylene biosynthesis in tobacco. *Plant Physiology* 95, 882–889.

Lee, E.H., Tingey, D.T. and Hogsett, W.E. (1988) Evaluation of ozone exposure indices in exposure response modelling. *Environmental Pollution* 53, 43–62.

Lefohn, A.S. (1992) The characterization of ambient ozone exposures. In: Lefohn, A.S. (ed.) *Surface-Level Ozone Exposures and Their Effects on Vegetation*. Lewis Publishing, Chelsea, Michigan, pp. 31–92.

Liebold, E. (1988) Wirkungsmodell der längerfristigen Schadensprognose im SO_2-geschädigten Fichtenwald. *Wissenschaftliche Zeitschrift der Technische Universität Dresden* 37, 243–247.

Lorenzini, G., Triolo, E. and Materazzi, A. (1984) Evidence of visible injury to crop species by ozone in Italy. *Rivista Ortoflorofrutticoltura Italiana* 68, 81–84.

Loye-Pilot, M.D., Martin, M.D. and Morelli, J. (1989) *Water Pollution Research Reports. EROS 2000*. European Commission, Brussels, pp. 368–376.

Matyssek, R., Günthardt-Goerg, M.S., Maurer, S. and Keller, T. (1995) Night time exposure to ozone reduces whole-plant production in *Betula pendula*. *Tree Physiology* 15, 159–165.

Mehlhorn, H. and Wellburn, A.R. (1987) Stress ethylene formation determines plant sensitivity to ozone. *Nature* 327, 417–418.

Millán, M.M., Artíñano, B., Alonso, L., Castro, M., Fernandez-Patier, R. and Goberna, J. (1992) *Meso-Meteorological Cycles of Air Pollution in the Iberian Peninsula*,

(MECAPIP). Air Pollution Research Report 44 (EUR N° 14834), European Commission DG XII/E-1, Brussels.

Millán, M.M., Salvador, R., Mantilla, E. and Artíñano, B. (1996) Meteorology and photochemical air pollution in southern Europe: experimental results from EC research projects. *Atmospheric Environment* 30, 1909–1924.

Millán, M.M., Salvador, R., Mantilla, E. and Kallos, G. (1997) Photo-oxidant dynamics in the western Mediterranean in summer: results from European research projects. *Journal of Geophysical Research* 102, 8811–8823.

Millán, M.M., Mantilla, E., Salvador, R., Carratalá, A., Sanz, M.J., Alonso, L., Gangoiti, G. and Navazo, M. (1999) Ozone cycles in the Western Mediterranean Basin: interpretation of Monitoring data in complex coastal terrain. *Journal of Applied Meteorology* (in press).

Miller, P.R. (1973) Oxidant-induced community change in a mixed conifer forest. *Advances in Chemistry Series* 122, 101–117.

Miller, P.R. and Millecan, A.A. (1971) Extent of oxidant air pollution damage to some pine and other conifers in California. *Plant Disease Reporter* 55, 555–559.

Miller, P.R. and Parmeter, J.R. (1967) Effects of ozone injury to ponderosa pine. *Phytopathology* 57, 822.

Miller, P.R., Parmeter, J.R., Flick, B.H. and Martinez, C.W. (1969) Ozone dosage response of Ponderosa pine seedlings. *Journal of the Air Pollution Control Association* 19, 435–438.

Minnocci, A., Paniucucci, A. and Vitagliano, C. (1995) Gas exchange and morphological stomatal parameters in olive plants exposed to ozone. In: Lorenzini, G. and Soldatini, G.F. (eds) *Responses of Plants to Air Pollution*. Agricoltura Mediterranea, special volume, pp. 77–81.

NABEL (1995) *Luftbelastung 1994*. Schriftenreihe Umwelt 244, Bundesamt für Umwelt, Wald und Landschaft, Bern.

Naveh, Z., Steinberger, E.H., Cahim, S. and Rotmann, A. (1980) Photochemical air pollutants. A threat to Mediterranean coniferous forest and upland ecosystems. *Environmental Conservation* 7, 301–309.

Orciari, R., Georgiadis, T., Fortezza, F., Alberti, L., Leoncini, G., Venieri, L., Gnani, V., Montanari, T. and Rambelli, E. (1999) Vertical evolution of photochemical ozone over greater Ravenna. *Annali di Chimica* (in press).

Patterson, M.T. and Rundel, P.W. (1995) Stand characteristics of ozone-stressed populations of *Pinus jeffreyi* (Pinaceae): extent, development, and physiological consequences of visible injury. *American Journal of Botany* 82, 150–158.

Prinz, B., Krause, G.H.M. and Stratmann, H. (1982) *Waldschäden in der Bundesrepublik Deutschland*, LIS-Berichte 28, Landesanstalt für Immissionsschutz des Landes Nordrhein-Westfalen, Essen.

Reich, P.B. (1983) Effects of low concentrations of O_3 on net photosynthesis, dark respiration, and chlorophyll contents in aging hybrid poplar leaves. *Plant Physiology* 73, 291–296.

Reinert, R., Gimeno, B.S., Selleras, J.M., Bermejo, V., Ochoa, M.J. and Tarruel, A. (1992) Ozone effects on watermelon plants at the Ebro delta (Spain). Symptomatology. *Agriculture, Ecosystems and Environment* 38, 41–49.

Richards B.L., Taylor, O.C. and Edmunds, G.F., Jr (1968) (Ozone needle mottle of pine in southern California. *Journal of Air Pollution Control Association* 18, 73–77.

Sanz, M.J. and Calatayud, V. (1999) An example of combined studies: Mesoscale ozone distribution and its effects on forest species (*Pinus halepenis* Mill.). Submitted to I Technical Workshop on Ozone Pollution in Southern Europe. Valencia, 1997 (in press).

Sanz, M.J. and Millán, M.M. (1998) The dynamics of polluted airmasses and ozone cycles in the western Mediterranean: relevance of forest ecosystems. *Chemosphere* 36, 1089–1094.

Sanz, M.J., Calatayud, V. and Calvo, E. (1998a) Spatial pattern of ozone injury in Aleppo pine and air pollution dynamics in the Mediterranean. *Annales Geophysicae* 16 (Suppl. III), 939.

Sanz, M.J., Carratala, A., Mantilla, E., Dieguez, J.J. and Millán, M. (1998b) Daily ozone patterns and AOT40 index on the east coast of the Iberian Peninsula. *Annales Geophysicae* 16 (Suppl. III), 939.

Schenone, G. (1993) Air pollution and vegetation in the Mediterranean area. *Mediterranea* 2, 49–52.

Schenone, G. and Lorenzini, G. (1992) Effects of regional air pollution on crops in Italy. *Agriculture, Ecosystems and Environment* 38, 51–59.

Schuh, G., Wildt, J., Rockel, P., Rudolph, J., Wolters, H.W. and Kley, D. (1996) Terpene and sesquiterpene emission rates from sunflower. In: *Proceedings of the Seventh European Symposium on Physicochemical Behaviour of Atmospheric Pollutants, Venice.* EUR 17482 EN, Commision of the European Community, Brussels, pp. 290–294.

Schulze, E.-D., Kelliher, F.M., Körner, C., Lloyd, J. and Leuning, R. (1994) Relationships among maximal stomatal conductance, ecosystem surface conductance, carbon assimilation rate, and plant nitrogen nutrition: a global ecology scaling exercise. *Annual Reviews of Ecology and Systematics* 25, 629–660.

Seufert, G. (1997) BEMA. A European Commission project on biogenic emissions in the Mediterranean area. *Atmospheric Environment* 31, 1–256.

Simini, M., Skelly, J.M., Davis, D.D., Savage, J.E. and Comrie, A.C. (1992) Sensitivity of four hardwood species to ambient ozone in north central Pennsylvania. *Canadian Journal of Forest Research* 22, 1789–1799.

Skelly, J.M., Innes, J.L., Savage, J.E., Snyder, K.R., van der Heyden, D., Zhang, J. and Sanz, M.-J. (1999) Observation and confirmation of foliar ozone symptoms of native plant species of Switzerland and Southern Spain. *Water, Soil and Air Pollution* (in press).

Thunis, P. and Cuvelier, C. (1997) *Report on the Preliminary BEMA Model Simulations. Using Data from Castelporziamo (Italy) and Burriana (Spain).* EUR 17291 EN, European Commission, Brussels.

Tiedemann, A.V. and Herrmann, J.V. (1992) First record of grapevine oxidant stipple in Germany and effects of field treatments with Ethylenediurea (EDU) and Benomyl on the disease. *Zeitschrift für Pflanzenkrankheiten und Pflanzenschutz* 99, 553.

Velissariou, D., Barnes, J.D., Davison, A.W., Pfirrmann, T., Maclean, D.C. and Holevas, C.D. (1992) Effects of air pollution on *Pinus halepensis* (Mill.): pollution levels in Attica, Greece. *Atmospheric Environment* 26A, 373–380.

Velissariou, D., Gimeno, B.S., Badiani, M., Fumgalli, I. and Davison, A.W. (1996) Records of ozone visible injury in the ECE Mediterranean region. In: Kärenlampi, L. and Skärby, L. (eds) *Critical Levels for Ozone in Europe: Testing and Finalizing the Concepts. UN-ECE Workshop Report.* University of Kuopio, Kuopio, pp. 343–350.

Weber, J.A., Tingey, D.T. and Andersen, C.P. (1994) Plant response to air pollution. In: Wilkinson, R.E. (ed.) *Plant–Environment Interactions.* Marcel Dekker, New York, pp. 357–389.

Wolfenden, J. and Mansfield, T.A. (1991) Physiological disturbances in plants caused by air pollutants. *Proceedings of the Royal Society of Edinburgh* 97B, 117–138.

Ziomas, I.C., Gryning, S.E. and Borsteing, R.D. (1998) The Mediterranean campaign of photochemical tracers – transport and chemical evolution (MEDCAPHOT – TRACE), Athens, Greece 1994–1995. *Atmospheric Environment* 32, 2043–2326.

Revitalization and Restoration of Boreal and Temperate Forests Damaged by Air Pollution

11

M.V. Kozlov,[1] E. Haukioja,[2] P. Niemelä,[2,3] E. Zvereva,[1] and M. Kytö[2]

[1]Section of Ecology, Department of Biology, University of Turku, Turku, Finland; [2]Vantaa Research Centre, Finnish Forest Research Institute, Vantaa, Finland; [3]Faculty of Forestry, University of Joensuu, Joensuu, Finland

The influence of industrial pollution on the health of boreal and temperate forests has long been recognized as a serious scientific and applied problem. Numerous field experiments have been established to determine practical measures for both the alleviation of pollution impacts on stand vitality and the rehabilitation of forest ecosystems. At incipient stages of forest damage, the impacts of acidic deposition and heavy metals can be mitigated by the application of limestone and other compensatory fertilizers and by the adoption of silvicultural measures which increase tree vigour. In clearcuts and areas of extensive tree mortality surrounding major sources of pollution, the regeneration of forests is hindered by the unfavourable abiotic environment and soil toxicity, which requires soil detoxification and other site preparation measures. Under conditions of continuous pollution impact, it is possible to reduce soil erosion in deforested habitats by planting native, metal-tolerant woody plants such as willows and birches. Both the vigour of stands impacted by pollution and the establishment of trees in heavily polluted sites can be enhanced by the removal or reduction of co-occurring environmental stressors such as wind and/or drought. Although a large number of studies have been carried out in the field of forest restoration, the large-scale practical application of these to boreal and temperate forests damaged by pollution is still restricted to a very few areas.

1   Introduction

Large areas of forest require rehabilitation as a result of the impacts of acidic
deposition and other forms of pollution (Downing *et al.*, 1993; Newson, 1995).
The problem is particularly apparent in Central Europe, with the most striking
example being the 'Black Triangle', an area along the German–Czech–Polish
border. This region has been heavily affected by industrial pollution over
the past 50 years (see Chapter 8, this volume), with severe consequences
for the forests, landscapes, environment and public health (Fanta, 1997).
Extensive forest mortality around large sources of pollution has sometimes
transformed forests to barren 'moonscapes', and has attracted considerable
public and scientific attention during the past decades (Winterhalder, 1984;
Bridges, 1991).

   The full restoration of forests differs from the restoration of most other
types of vegetation, mostly in relation to the time required to complete the process (Ashby, 1987), and there is no known example of a successfully completed
restoration (*sensu stricto*) of the forest communities on barren lands contaminated by industrial pollution. The land reclamation programme in Sudbury,
Canada (Gunn, 1995; Chapter 7, this volume), the only practical example of a
large-scale restoration of an area impacted by extreme pollution conditions,
has only been in operation for a relatively short period compared to the time
required for successional processes to form forest ecosystems (Winterhalder,
1995). However, the numerous examples of the reclamation of forests affected
by chronic acidification (Latocha, 1990; Thomasius, 1990; Katzensteiner
et al., 1992; Lenz and Haber, 1992; Hüttl and Zöttl, 1993; Kubelka *et al.*, 1993;
Tesar, 1994; Fanta, 1997; Mitsch and Mander, 1997; Pawlowski, 1997)
provide useful information for the further implementation of restoration
measures.

   In this chapter, we concentrate on: (i) the revitalization of forests slightly
to moderately damaged by air pollution; and (ii) the restoration of forests in
deforested areas subjected to pollution, with specific attention being paid to
industrial barrens adjacent to major pollution sources, where contamination-
induced environmental changes have prevented or significantly delayed the
regeneration of forest plants.

2   Chemical revitalization of damaged forests

Air pollutants and acidic precipitation affect the foliar tissue of forest trees by
enhancing the leaching of nutrient ions (K, Mg and Ca) from the canopy
(Schulze, 1989; Katzensteiner *et al.*, 1992; Hüttl and Hunter, 1992; Hüttl,
1993b) or by causing nutritional disturbances in some other way (Hüttl,
1988a; Hüttl and Mehne, 1988). At the same time, soil acidification promotes
nutrient leaching from the upper soil horizons. Another important factor is

increased N input; nutrient imbalances may be induced in cases where the N nutrition is shifted from NO_3-N to NH_4-N (Hüttl, 1988b).

The intensity of the leaching of nutrients from both the canopy and the soil in comparison to the rate of supply of Mg, K and Ca by weathering determines whether a tree can maintain its nutritional status. Magnesium deficiency is therefore most likely on substrates poor in plant-available Mg, and the poor availability of this nutrient was largely responsible for the decline of trees in areas such as the Black Forest of Germany during the late 1970s and early 1980s (Hüttl, 1988a, b; Hüttl and Mehne, 1988). Current revitalization strategies therefore require: (i) the identification of nutritional disturbances; and (ii) the application of diagnostic fertilization techniques to alleviate these disturbances (Zöttl and Hüttl, 1986; Hüttl, 1988a, 1989; Hüttl and Fink, 1988; Liu and Trüby, 1989; Insam et al., 1994; Katzensteiner et al., 1995).

2.1 Liming

Effects on existing stands

The main objective of older liming trials and practices, which were sometimes combined with phosphorus fertilization, was to ameliorate anthropogenically disturbed topsoil layers (Hausser, 1958; Mitscherlich and Wittich, 1958). On some soils, liming is necessary to counteract calcium depletion caused by whole-tree harvesting (Erich and Trusty, 1997). In many studies, it was expected that an increase in soil Ca concentrations would enhance the mobilization of nutritional elements in acidic forest soils and thus increase stand productivity (Zöttl and Hüttl, 1986). However, the majority of German liming trials indicate either insignificant or no growth increase (Aldinger, 1987; Zöttl, 1990; Evers and Hüttl, 1990). Conversely, in both Finland and Sweden, liming of Norway spruce and Scots pine (*Pinus sylvestris* L.) stands resulted in slight growth reductions (Derome et al., 1986; Andersson and Persson, 1988).

As expectations of increased tree growth were not met, and important questions about the ecological side-effects of liming arose (see below), widespread liming was discontinued in the mid-1970s (Hüttl and Zöttl, 1993). However, with the appearance of widespread symptoms of magnesium deficiency (Hüttl and Mehne, 1988), liming was reconsidered as a means of stabilizing stands affected by acidic deposition (Tveite et al., 1991; Hüttl and Zöttl, 1993) and improving surface water quality (Nyström et al., 1995). Over the past two decades, dolomite lime or easily soluble neutral salts such as magnesium sulphate have been applied to many forests in central and northern Europe (Ulrich et al., 1979; Ulrich, 1981; Kaupenjohann and Zech, 1989; Liu and Ende, 1989; Hüttl, 1991; Kaupenjohann, 1991; Klimo, 1991), the United Kingdom (Howells and Dalziel, 1992) and the USA (Gubalba and Driscoll, 1991). A notable exception is Switzerland, where all forms of forest fertilization, including liming, are forbidden by law because of the supposed adverse ecological impacts.

Diagnostic liming trials conducted in Germany suggest that Mg-containing lime improves the Mg status of soils, although the effect is considerably slower compared with rapidly soluble Mg fertilizers (Hüttl and Zöttl, 1993). Some of the trials conducted in the Bohemian Mountains have demonstrated a positive effect of Mg-containing lime on the growth of Norway spruce (Katzensteiner et al., 1995). More recently, the application of granulated magnesium-rich limestone (5% Mg) and boron to Scots pine stands growing in south-west Finland on soils heavily contaminated with Ni and Cu enhanced both the above-ground volume increment and the fine-root production (Helmisaari et al., 1999; Mälkönen et al., 1999). This suggests that biomass production on metal-contaminated soil is limited by macronutrient deficiencies caused by the replacement of base cations by heavy metals, retarded decomposition and mineralization of the litter, and diminished nutrient uptake caused by disturbances to the functioning of the fine roots and mycorrhizae (Helmisaari et al., 1995; Mälkönen et al., 1999). However, the stimulation of fine-root growth in the uppermost soil layers following lime application may not always be beneficial to the stand as it increases the risk of drought, frost and windthrow damage (Hildebrand, 1990).

Effects on soil chemistry and element fluxes

Forest soil acidification is a natural phenomenon over large forest areas. However, the process can be accelerated by various forest/land use management practices and by acidic deposition (Hüttl, 1993b). Liming can increase the pH and base saturation and reduce the exchangeable Al concentrations in the soil organic layer (Derome et al., 1986; Marschner et al., 1989; Derome and Lindroos, 1998; Nikonov et al., 1999). It can also greatly increase the exchangeable Ca and Mg concentrations (Hüttl and Zöttl, 1993; Kalatskaya, 1997; Nikonov et al., 1999). Furthermore, liming of metal-contaminated soils can markedly decrease the Cu and Ni concentrations in the organic layer, presumably due to the immobilization of metals through precipitation or absorption (Winterhalder, 1988; Mälkönen et al., 1999).

An increase in the retention of heavy metals in the forest floor demonstrates the effectiveness of the forest as a filter for anthropogenic pollutants, but also increases the risk of adverse effects on roots and microorganisms (Marschner et al., 1992). Lime application may at least temporarily lead to 'acidification pushes' in deeper soil horizons (probably due to exchange processes) (Evers and Hüttl, 1990) and may also cause other side-effects such as increased NO_3^- leaching combined with an equivalent loss of cations such as Mg, Ca and K (Hüttl and Zöttl, 1993). The surface application of lime greatly enhances internal acid production from nitrification in the forest floor: this internal acid production surpasses external inputs by a factor of three to five (Marschner *et al.*, 1992).

Liming may also cause increased nitrate concentrations in drainage water, the decomposition of surface humus, an increase in organic Cu complexes due to the mobilization of water-soluble humus decomposition products (Kreutzer, 1986, 1995; Wenzel and Ulrich, 1988; Marschner, 1990), and an increase in sulphate concentrations in the soil solution (Mälkönen *et al.*, 1999). The latter effects may be linked to either an increase in Ca and Mg concentrations, because sulphate acts as the accompanying anion in the leaching of these cations from the soil profile (Bergseth, 1978), or an increase in sulphate release rates due to enhanced microbial activity (Marschner, 1993). Liming with high doses of $CaCO_3$ has reduced the pools of soil carbon in relation to unlimed areas in Norway spruce and beech (*Fagus sylvatica* L.) forests in southern Sweden (Persson *et al.*, 1995) as well as in both hardwood and softwood stands in North America (Erich and Trusty, 1997). The effects of liming in some soil types may still be detectable 40 years after the treatment (Persson *et al.*, 1995), suggesting that ecological risks may increase in the future due to the expected increase in nitrogen deposition (Kreutzer, 1986; Kreutzer *et al.*, 1989; Matzner and Meiwes, 1990; Kreutzer and Schierl, 1992).

In conclusion, liming in forests should not be a general practice, but should be used carefully under specific site conditions (Kubelka *et al.*, 1993; Kreutzer, 1995). This requires balancing the anticipated beneficial and adverse effects; in particular, sites with subsurface and surface flow contaminated by heavy metals and sites with low humus content should be excluded from liming (Hüttl and Zöttl, 1993; Kreutzer, 1995). However, the liming of suitable soils will help to compensate for potential future acidic deposition (Hüttl, 1989) and will result in the long-term beneficial alteration of the chemical status of acidified soils (Hüttl and Zöttl, 1993).

Effects on litter-decomposing organisms

Liming induces great changes in the forest soil biota, even in situations where liming is applied to counteract the effects of acidic deposition (Kreutzer, 1995). Liming can significantly increase bacterial biomass (Zelles *et al.*, 1990) and microbial activity (measured as increased soil respiration) in metal-contaminated soils (Smolander *et al.*, 1994; Fritze *et al.*, 1996). The number of heterotrophic bacteria increased by a factor of 50 after liming, and autotrophic ammonium and nitrite oxidizers were found exclusively on limed plots (Papen *et al.*, 1991). However, in another liming experiment the observed reduction in net N mineralization during the first year after liming was explained by microbial immobilization (Marschner, 1993); this reduction caused the N limitation of decomposers despite the relatively high N inputs (Marschner *et al.*, 1992).

The total abundance (or biomass) of *Collembola* and litter-decomposing macrofungi did not respond to liming, but species composition and community structure were markedly changed in both these groups (Agerer, 1991), as were nematode communities (DeGoede and Dekker, 1993). The abundance of earthworms was greatly increased in limed plots (Makeschin, 1991; Kreutzer, 1995). Thus, as with soil chemistry, the negative side-effects on the soil biota may outweigh any advantages to be gained from liming. This particularly concerns protected ecosystems endangered by acidic deposition: the application of lime may change them more than acidification would (Schornick, 1989; Kreutzer, 1995) and may even lead to the extinction of some species (Ahrens, 1994).

2.2 Acidification

Soil pH may reach very high values (9–10) near enterprises releasing alkaline dust. For example, industrial barrens around the 'Magnesit' factory at Satka, Ural region, covered 20 km^2 in 1980. Soil in these barrens was covered by a layer of dust primarily containing magnesium oxide. Recovery of the barren areas was considered impossible prior to a decline in dust emissions, whereas revitalization of forests in areas with intermediate and low pollution required application of nitric (10%) and hydrochloric (5%) acids in very high amounts (0.7–1.3 kg m^{-2}). Peat and sawdust were also used to acidify the topsoil. After the decrease in soil pH, NPK fertilization had a positive effect on the growth and vigour of birch and Scots pine (Luganskij *et al.*, 1983).

2.3 Fertilization

Mitigation of macronutrient deficiency

Fertilization has long been used in forestry both to improve vitality and to enhance the productivity of stands (Smolyaninov, 1969; Kukkola and Saramäki, 1983), including forests suffering from sulphur dioxide and fluorine-containing emissions (Stchavrovskij *et al.*, 1983; Golutvina *et al.*, 1987). As many forest ecosystems have developed under conditions in which nitrogen supply is the limiting factor for tree growth, N-containing mineral fertilizers were widely applied between 1950 and the 1970s (Shumakov, 1977; Hüttl, 1991; Katzensteiner *et al.*, 1995). They were also used to aid the recovery of degraded ecosystems (Baule and Fricker, 1967; Fiedler *et al.*, 1973). The effects of fertilization in polluted forests may last for 10 or more years (Stchavrovskij *et al.*, 1983; Golutvina *et al.*, 1987).

A change has occurred in recent decades, as nitrogen inputs in polluted areas now exceed the demand of some forest ecosystems. As a result of high

N-deposition, elements such as Mg have become limiting factors for tree growth on acidic substrates (Katzensteiner *et al.*, 1995). In such circumstances, the application of rapidly soluble mineral NPK-fertilizers can have adverse effects on tree growth (Golutvina *et al.*, 1987) and nutrition and may lead to NO_3^- contamination of seepage water, particularly under conditions of N-saturation (Katzensteiner *et al.*, 1995). For example, the application of both mineral and organic fertilizers in a severely damaged Norway spruce stand in the Bohemian Massif (Austria) resulted in a highly elevated content of NH_4^- and NO_3^- in the soil water. Effects on respiration, litter decomposition and N mineralization were, however, weak (Insam *et al.*, 1994).

The application of mineral NPK fertilizers to the industrial barrens around the Severonikel copper–nickel smelter resulted in rapid mortality of birch seedlings (Kozlov and Haukioja, 1999) and a significant decline in vitality of mature birch trees (M. Kozlov, unpublished data), probably due to the phosphate-enhanced toxicity of nickel-contaminated soils (Crooke and Inkson, 1955). In contrast, the application of fertilizers to similar metal-contaminated habitats at Sudbury, Canada, had quite a positive effect (Winterhalder, 1983, 1988), presumably because metal toxicity at Sudbury was decreased by the simultaneous application of limestone (Lautenbach *et al.*, 1995). The application of mineral NPK fertilizers in acidified stands, even those with a high soil Mg content, induced Mg deficiencies and led to nutritional imbalances in Norway spruce. Conversely, an organic slow-release fertilizer amended with magnesite-derived fertilizers led to balanced nutrition and a fast recovery of tree health. In this situation, increased rates of NO_3^- leaching were recorded for only the 3 years following fertilization (Katzensteiner *et al.*, 1995).

Following the widespread appearance of Mg deficiency in Europe in the 1980s (Hüttl, 1993a), a large number of diagnostic Mg fertilizer trials were established (Hüttl and Fink, 1988; Hüttl, 1989, 1991; Zöttl, 1990; Hüttl and Hunter, 1992; Katzensteiner *et al.*, 1992, 1994, 1995; Hüttl, 1995; Schaaff, 1995; Lukina and Nikonov, 1998; Mälkönen *et al.*, 1999; Nikonov *et al.*, 1999). These trials demonstrated that Mg deficiency in Norway spruce, silver fir (*Abies alba* Miller), Scots pine, Douglas fir (*Pseudotsuga menziesii* (Mirbel) Franco) and beech (*Fagus sylvatica* L.) can be corrected through the application of soluble Mg fertilizers. Site- and stand-specific K and Mg fertilization led to the successful recovery of affected deciduous and coniferous stands of all ages and resulted in a long-lasting improvement in soil nutritional status, above-ground biomass production and fine-root vitality (Ranft, 1982; Hüttl, 1988b, 1995; Schaaff, 1995; Mälkönen *et al.*, 1999).

Although nearly any source of Mg is able to improve the Mg nutrition of trees, in some cases an improvement in stand vitality and growth was recorded only following the application of dolomitic lime or magnesite; for example, the fertilization of forests on acidified soils was particularly efficient when the supply of nutrients was combined with pH-stabilization measures (Katzensteiner *et al.*, 1995). A combination of a slow-release organic fertilizer and a magnesite fertilizer on sites with Mg-deficiency symptoms is better than a single magnesite fertilization and restores tree vigour within a shorter timespan (Katzensteiner *et al.*, 1992).

The above studies indicate that, as with liming, the potential adverse effects of fertilizer applications can easily be minimized when the correct fertilizers are used in the proper amounts at the appropriate times on specific sites (Hüttl, 1989).

Amelioration of micronutrient deficiencies

Micronutrient deficits can have drastic effects on plant vitality, and some micronutrients such as boron, zinc, copper, iron and manganese have been applied in combination with macronutrients in some of fertilization trials in polluted forests (Golutvina *et al.*, 1987; Mälkönen *et al.*, 1999). However, the contribution of these micronutrients to an overall positive effect of fertilization (e.g. Päivinen, 1994) has not, to our knowledge, been confirmed in specifically designed experiments.

2.4   Stimulation of plant growth

An experimental application of a growth stimulator (phosphorylated benzimidazole) used in agriculture to deciduous plants in polluted forests resulted in an increase in the height increment of birch by 10–20% (Alexeyev *et al.*, 1985). However, this method does not seem to have been further developed.

3   Silvicultural management of damaged forests

The health of forests depends not only on soil nutritional quality, but also on a number of other environmental factors, including climate, water supply, stand density and composition, weeds, pests and pathogens. Therefore, the direct mitigation of pollution effects is only one of several options; any technically possible and economically feasible measure which would improve forest growth could be used to maintain and improve the stability of forests damaged by pollution (Thomasius, 1990).

3.1   Industrial approaches

Efforts to prevent losses of commercial timber in stands damaged by pollution in the Bohemian part of the Ore Mountains led to clearcuts over large areas. The first complex strategy for forest management in affected areas (developed in 1957) was based on traditional spruce monocultures, with the substitution of Norway spruce by introduced spruce species considered to be more tolerant of pollution than Norway spruce. Colorado blue spruce (*Picea pungens* Engelm.) initially appeared to be the most successful in damaged habitats, but the afforestation efforts were unsuccessful in the long term (Tesar and Tichy, 1990). A gradual increase in the appreciation of the importance of naturally regenerated groups of autochthonous species such as birch (*Betula* spp.) and mountain ash (*Sorbus aucuparia* L.), for the maintenance of forest stability, along with the failure of some introduced spruce species, has resulted in a shift in emphasis from 'technical' restoration to 'ecological' restoration (Tesar and Tichy, 1990).

3.2 Ecological approaches

Sheltering

The large-scale clearcutting of damaged forests carries great ecological risks, especially in mountains where the forest has irreplaceable soil and climate protective roles. Such forests should not be felled until adequate regeneration has been established (Latocha, 1990; Tesar and Tichy, 1990). Consistent with this conclusion, the development of industrial barrens around the major sources of pollution in the northern boreal forest zone have always been initiated by some catastrophic event, such as clear-felling or fire (Kozlov, 1997). If cutting is considered as the first stage in the process of forest regeneration, whenever possible no large uninterrupted clearcuts should be made (Kubelka et al., 1993). Cuts of 15–30 m strips have been recommended as facilitating the reconstruction of damaged forests. The strips of remaining forest can be removed after the establishment of regeneration in the cleared areas (probably after 7–10 years in Upper Silesia) (Latocha, 1990).

Thinning

Stand density is linked to tree vitality via competition with neighbours, resistance to wind and snow, and the ability to regulate microclimate. It has been suggested that young stands in areas affected by acidic deposition should be established with a low density which would allow plants to develop large symmetrical crowns, whereas higher densities should be maintained in older stands in order to increase their resistance (Thomasius, 1990). However, this conflicts with information obtained in the Bohemian Mountains, where the heavy thinning of a 15-year-old Norway spruce stand prevented further degradation, while the vitality of both the control and slightly thinned stands markedly decreased (Slodicák, 1994).

Conversion

It has been suggested that forest stands which are unable to fulfil their production, protection or recreation functions as a result of recent or expected damage by pollution should be gradually converted (e.g. Thomasius, 1990). The choice of tree species for such a conversion depends primarily on the site conditions (Latocha, 1990; Tesar and Tichy, 1990; Thomasius, 1990; Kubelka *et al.*, 1993). Basic principles, such as the use of indigenous tree species, especially early successional species such as birch, aspen (*Populus tremula* L.) and mountain ash, and the formation of multi-species assemblages instead of monocultures, seem to be important (Latocha, 1990; Thomasius, 1990; Kubelka *et al.*, 1993; Tesar, 1994; Fanta and Kobus, 1995). An important goal is to allow the development of a forest with a tree composition typical for the region (Först and Kreutzer, 1977) and seed material should preferably represent a local ecotype (Luganskij *et al.*, 1983; Kubelka *et al.*, 1993).

Within-species variation in resistance to pollutants has been discovered in several woody plants (Pietelka, 1988) and (at least in some species) this variation is heritable (Rachwal and Wit-Rzepka, 1989; Utriainen *et al.*, 1997). Therefore, the individuals of relatively sensitive tree species (Norway spruce, Scots pine) surviving on contaminated sites may represent genotypes resistant to pollution. This hypothesis was confirmed experimentally in the early 1980s by sowing seeds of different origin in a site polluted by alkaline dust. Seeds of Scots pine and birch collected from the polluted area showed better results than seeds from an unpolluted forest (Luganskij *et al.*, 1983). Although there is an obvious need to study these resistant individuals (Kubelka *et al.*, 1993), this does not appear to have been done. This is an important omission as recent developments in tree breeding may enable the future development of resistant conifer genotypes for planting in toxic soils.

4   Restoration of damaged forests

The regeneration procedure that is adopted obviously depends on the scale of the disturbance and the site conditions, primarily the microclimate, soil toxicity and soil nutritional quality. If neighbouring stands ameliorate the microclimate sufficiently, target forest stands with the original or an adjusted species composition can be established directly. On large clearcuts and on extensive industrial barrens, where the forest microclimate has been severely disrupted, substitute forest stands should first be established followed by the target ones under the protection of the former (Kubelka *et al.*, 1993). In the most severely affected areas, where soil toxicity inhibits the growth of seedlings (Whitby *et al.*, 1976; Winterhalder, 1984, this volume), soil detoxification by liming should first be undertaken. The establishment of a herbaceous grass cover is desirable and may be necessary before trees and shrubs can be successfully established (Winterhalder, 1984, 1988). However, regardless of site conditions, monospecific stands should not be established, as these are unstable even under normal conditions and they are even more so in regions with high pollution impacts (Kubelka *et al.*, 1993).

4.1 Site preparation

Mechanical preparation

Different methods of mechanical preparation have been used: (i) to enhance the mechanical soil quality; (ii) to improve the soil water regime; (iii) to create microhabitats suitable for tree establishment; (iv) to remove any continuous grass cover; and (v) to remove the heavily polluted topsoil. An important consideration when choosing the method of mechanical soil preparation is the prevention of soil erosion (Luganskij *et al.*, 1983; Kubelka *et al.*, 1993). The need for such measures is illustrated by the extensive soil erosion in the industrial barrens around the copper–nickel smelters in the Kola Peninsula, which has resulted in the exposure of a residual stony surface or bedrock over up to half of the surface area. In some field trials, stones have been removed from experimental plots (Lukina *et al.*, 1997), but this is labour-intensive and usually impractical.

Soil ploughing after clearcutting is a common forestry practice which has positive effects on tree growth in the first phases of forest regeneration (Valtanen and Engberg, 1987), including in polluted areas (Luganskij *et al.*, 1983). In the Upper Silesian industrial region, full tillage of sandy soil promoted better growth and development of nearly all tree species in their juvenile period than other methods of soil preparation (in furrows, disks, etc.). Full tillage resulted in a decrease in the soil acidity, reduced metal (at least for Zn and Pb) contents, enhanced microbiological activity and decreased root infections of young trees by root-rot fungi (*Heterobasidion annosum* (Fr.) Karst.). Conversely, mycorrhizal infestation of Scots pine roots was lower in ploughed areas (Väre, 1988), and the soil content of N, K and Mg in the ploughed layer also decreased, requiring compensatory measures (Latocha, 1990). The exposure to acidic deposition of deeper soil layers containing bound heavy metals may increase the bioavailability of some toxic elements such as zinc and mercury (Castrén *et al.*, 1990; Nuorteva, 1990). Tilling also slightly increased the proportion of damaged leaves in birches, possibly because the vigorously growing plants were more attractive to herbivores (Nuorteva, 1990). Strips scraped perpendicular to the prevailing wind direction and separated by mounds 1.2–2.5 m high are more suitable for plant establishment than a flat soil surface, due to both higher habitat diversity and the protection (primarily from wind) offered by the mounds. Mounds are particularly important in large barren areas unprotected by neighbouring stands (Kubelka *et al.*, 1993; Pysek, 1994).

Deforested areas in the Czech mountains are frequently colonized by the grass *Calamagrostis villosa* (Chaix) J. Gmelin, which makes the replanting of forest trees extremely difficult or even impossible. Attempts to remove the topsoil by bulldozing areas covered by *C. villosa* created plots with mineral soils of low fertility and no seedbank. These plots were first colonized by *C. villosa* but after about 5 years wavy hair-grass (*Deschampsia flexuosa* (L.) Trin.) became the dominant species. This replacement was associated with changes in several soil features (particularly a decrease in soil acidity and an increase in Ca, K and other nutrients) and is considered to be advantageous in relation to habitat restoration, as it promoted the establishment of trees (birch, mountain ash) and therefore accelerated the natural succession leading to the establishment of a full forest cover (Pysek, 1994).

One suggestion has been to rely on spontaneous vegetation development in sites where the topsoil has been removed, an option favoured because of its low cost and the natural selection of woody species tolerant to pollution (Pysek, 1992). However, experience gained during the reclamation of mined lands has shown that soil development on sites missing an organic soil layer can be assisted and accelerated by the introduction of plant species most appropriate to the restoration of those sites, although natural ecosystem development can then be left to occur without further assistance (Bradshaw, 1997).

Chemical preparation

The reclamation of the industrial barrens at Sudbury, Canada, which were characterized by heavily contaminated acidic soils, started with very large (10 t ha^{-1}) applications of lime, applied in early spring. Later in the summer, large (400 kg ha^{-1}) applications of phosphorus fertilizer were made (Lautenbach *et al.*, 1995). No adverse effects of these treatments have been reported to date. They not only improved the soil conditions, but also improved the quality of the drainage water from the treatment areas (Skraba, 1989). Consequently, the liming contributed to an increased pH and alkalinity in two local lakes (Yan *et al.*, 1995).

Around the Severonikel and Pechenganikel smelters in the Kola Peninsula, Russia, experimental applications of dolomitic limestone (2 t ha^{-1}) plus NPK fertilizer (750 kg ha^{-1}) also yielded better results (Lukina *et al.*, 1997) than other treatments. Application of an organic fertilizer (bio-humus, i.e. horse manure converted by earthworms) also resulted in grass establishment, but the applications required to achieve this result (10–30 t ha^{-1}: Lukina *et al.*, 1997) are unrealistic for large-scale reclamation programmes.

During the reforestation of clearcuts exposed to acidic deposition, diagnostic fertilization and liming were applied in the same way as for the revitalization of damaged stands (see above). Current recommendations are that liming be conducted at least twice: before the mechanical preparation of soils and after planting of seedlings (Kubelka *et al.*, 1993). Fertilization should be restricted to planting holes or planting rows in order to minimize competition from weeds (Thomasius, 1990). Herbicide application may enhance seedling establishment in habitats covered by *C. villosa* (Thomasius, 1990) or common couch (*Elytrigia repens* (L.) Desv. ex Nevski) (Luganskij *et al.*, 1983), but others recommend that herbicides are avoided during site preparation in polluted regions (Kubelka *et al.*, 1993).

Bioremediation

An alternative to the leaching or binding of toxic metals is phytoextraction, a technique that is sufficiently developed to be practical for large-scale applications. The method is cost-effective and involves the establishment of metal-accumulating plants and then the removal of phytomass containing metal contaminants (Cunningham *et al.*, 1995). The key components of phytoextraction are: (i) bioavailability of the toxic elements; (ii) efficient uptake by plant roots; and (iii) translocation to shoots which can easily be harvested. Both metal uptake and accumulation of metals in plant shoots can be enhanced through the application of synthetic chelates to the soil (Blaylock *et al.*, 1997). Partial improvement of site conditions in polluted habitats may also be achieved by sowing nitrogen-fixing plants such as *Lupinus* spp. (Golutvina *et al.*, 1987).

Work is being conducted to screen tree species for their ability to tolerate, take up, translocate, sequester and degrade organic compounds and heavy metal ions (Stomp *et al.*, 1993). Generally, woody plants are less effective than grasses and herbaceous species, but pioneer species such as birch seem to have high uptake rates (Eltrop *et al.*, 1991). Genetic engineering may significantly improve plant suitability for phytoremediation (Raskin, 1996). Although phytoremediation methods are now commercially available, their value for the rehabilitation of forest ecosystems remains uncertain. Phytoextraction is being increasingly used to detoxify 'problem' soils, such as those near landfill areas (e.g. Lee and Woo, 1998). A potential problem concerns the use of wood from such sites, but it is generally easier to recover the heavy metals from, for example, wood used as a source of bioenergy than it is to recover heavy metals in soils.

The recent use of soil-decomposing animals in ecosystem recovery seems to have limited application. The introduction of soil animals may represent an important step in the creation of a functioning decomposition system in rehabilitated soils. These animals may enhance the activity of microbial decomposers and improve mechanical soil quality. However, since they have a greater effect on the distribution of pollutants than on their concentrations, there is a risk that toxic elements will be transferred from deeper soil layers to the soil surface and subsequently into above-ground food webs.

4.2   Establishment of plant cover

Industrial barrens with low pollution loads

One of the main reclamation goals of the municipal land restoration programme at Sudbury (initiated in 1974) was to re-establish a forest similar to that which once covered the barren hills. This goal had two components: to conduct liming and initiate a grass–herb cover, and to introduce any tree and shrub species that did not colonize spontaneously (Lautenbach *et al.*, 1995). Between 1978 and 1993, 3070 ha of barren land was limed, fertilized and seeded with a mixture of grasses and nitrogen-fixing legumes (at a rate of 45 kg ha^{-1}) (Lautenbach *et al.*, 1995). The grass–legume establishment was rapidly followed by spontaneous colonization by birch, poplar and willows (Winterhalder, 1995). However, no coniferous species appeared in the first few years after treatment, although tests conducted in 1978–1982 demonstrated good growth and survival of conifers planted on reclaimed sites. Since 1983, about 1.7 million trees have been planted on previously grassed sites. Planting stock mainly consists of 2- or 3-year-old bare-root seedlings and 6-month- to 1-year-old containerized or paper-pot seedlings. The general aim has been to plant species typical of mature northern Ontario forests in order to accelerate the slow progression through successional birch and poplar woodland stages. Therefore, more than three-quarters of the planted material has been conifers, with an emphasis on pines (red pine, *Pinus resinosa* Ait.; white pine, *P. strobus*; and jack pine, *P. banksiana* Lamb.). As a result of the reclamation, the physical appearance of the landscape around Sudbury has been completely changed, at least within the public view corridors (Lautenbach *et al.*, 1995). See Chapter 7 for more details.

Industrial barrens with high pollution loads

In contrast to the Sudbury area, where the reclamation programme started following major reductions in the local emissions, pollution levels over much of the Kola Peninsula still exceed the critical loads (Moiseenko, 1994). There are no realistic prospects for rapid reductions in pollution levels (Chapter 4, this volume). Therefore, the reforestation programme developed by the Arkhangelsk Institute of Forests and Forest Chemistry (Tsvetkov, 1987; Tsvetkov and Chekrisov, 1987), which requires both emission reductions and high

financial costs, will be very difficult to implement. The most realistic reclamation goal for this particular area is to reduce the rate of vegetation decline and
soil erosion in the industrial barrens under the existing levels of pollution. This
may be achieved by the partial revegetation of industrial barrens with those
native woody plants that are most tolerant of the pollutant conditions (Kozlov
and Haukioja, 1999).

Grass sowing in barren habitats was successful only after chemical site
preparation (Evdokimova, 1995; Lukina *et al.*, 1997). However, some woody
plants can establish on barren sites without applications of lime and fertilizers. Field experiments with cuttings of boreal willow (*Salix borealis* (Fries.)
Nasar.) have demonstrated that this species can be used for partial
revegetation of industrial barrens even at high emission levels. Cuttings have
the best chance to establish if they are collected from female genets at a
clean site, grown for the first year in pots with fertile soil at an unpolluted
site, and planted in the industrial barrens early the next growing season
(Kozlov *et al.*, 1999). Three- to five-year-old seedlings of mountain birch
(*Betula pubescens* subsp. *czerepanovi* (Orlova) Hämet-Ahti) can tolerate the
harsh environmental conditions of the industrial barrens. The survival of
the seedlings can be improved by sheltering; watering contributed to the
better establishment of replanted seedlings, but once the first 3 weeks have
elapsed following re-planting, sheltered seedlings seem to survive under the
natural precipitation regime even in dry summers (Kozlov and Haukioja,
1999).

Thus, the pollution-mediated erosion of heavily degraded barren soils can
be controlled by artificial revegetation with native plant species even at the
existing (high) levels of pollution. Mountain birch and boreal willow are
suitable candidates for the reforestation of denuded landscapes. Removal of
stressors other than pollution (wind, drought, nutrient deficiency) helps plants
to establish in heavily polluted habitats in the region. In contrast to the earlier
programme (Tsvetkov, 1987; Tsvetkov and Chekrisov, 1987), the suggested
treatments are inexpensive and can even be used by volunteers participating
in municipal re-greening programmes (as done at Sudbury, see Chapter 7, this
volume).

Clearcuts

The reforestation measures to be adopted in clearcuts subjected to acidic
deposition do not in principle differ from those usually applied under normal
forestry practices (Luganskij *et al.*, 1983). However, container-grown plants
(normally with a bigger root-ball volume) are preferable to bare-root stock
because they increase the chances of successful establishment. Direct sowing is
mostly used for birch and mountain ash, provided that conditions permit
(Kubelka *et al.*, 1993).

5   Weeds and pests in relation to restoration problems

One of the causes of failure for reforestation programmes is competition between weeds and the trees. Seedlings should be protected until they over-grow weeds by: (i) chemicals (only those which are not accumulated in soils are suitable); (ii) mowing or crushing the weeds manually or mechanically; (iii) protection by mulching (Luganskij *et al.*, 1983; Kubelka *et al.*, 1993); or (iv) the use of tree shelters (Potter, 1991). However, we are not aware of any studies linking pollution loads and the impact of weed competition to the estab-lishment of woody plants.

Disintegration of forest stands as a result of pollution can be followed by the appearance of a number of insect pests, including species which have not previously been regarded as pests (Kubelka *et al.*, 1993; Zvereva *et al.*, 1997), and may require additional plant protection measures. This is an extremely complex area and the ecosystem relationships that are involved are poorly understood (see, for example, Chapter 9, this volume). In some situations, insect attacks can cause the failure of revegetation, especially under stressful environmental conditions (Louda, 1995).

Some silvicultural methods, such as the use of narrow clearcuts separated by belts of undisturbed stands, may decrease plant damage, due to the presence of viable populations of non-specific predators such as spiders and ants (Latocha, 1990). On the other hand, some measures used to revitalize declining forests may benefit insect herbivores. For example, theories dealing with tree resistance against herbivores and pathogens are usually based on the assumption that there is a limited amount of carbon to be allocated to different functions and tree parts (Herms and Mattson, 1992; Mattysek and Innes, 1999). Consequently, under moderate nutrient deficiency, fertilization should lead to reduced resistance if the carbon allocation is shifted from defence to growth (Bryant *et al.*, 1983; Lorio, 1986). This expectation was met in experi-ments conducted near the pulp and paper mill in Ust-Ilimsk, Siberia, where infestation of *Caragana arborescens* Lam. by midges increased by a factor of 16 and infestation of *Lonicera tatarica* L. by aphids increased by a factor of ten after fertilization (Golutvina *et al.*, 1988). However, experiments conducted around the Harjavalta copper–nickel smelter in south-west Finland did not reveal any effects of either fertilization or liming on the resistance of Scots pine to bark beetles of the genus *Tomicus* (Kytö *et al.*, 1998). The lack of fertilization effects on Scots pine resistance is consistent with the conclusion by Haukioja *et al.* (1998) that the predictions of the growth-differentiation hypothesis seem to be valid for condensed tannins only, and not for terpenoids (which are mainly responsible for the resistance of pines) or hydrolized tannins.

The damage caused by red deer (*Cervus elaphus* L.) browsing in Central Europe is unacceptably high (Kubelka *et al.*, 1993); the costs of protecting young trees against deer browsing are higher than those of establishing a new forest (Fanta, 1997). Traditional methods to protect young trees are impractical on large (often more than 100 ha) clearcuts and modern methods have been ineffective. The only solution is to reduce the number of game (Tesar and Tichy, 1990; Kubelka *et al.*, 1993). However, as with weeds, the interactions between deer browsing and pollution remain unknown.

6 Conclusions

The reduction of emissions is a critical requirement for any reclamation programme. However, until measures to reduce air pollution are fully realized, efforts should be devoted to improving the vitality and maintaining the stability of affected ecosystems (Hüttl, 1989; Nikonov *et al.*, 1999). Summarizing the situation in several forested areas of Central Europe, Lenz and Haber (1992) concluded that changes to ecosystem structure are difficult to rectify as essential processes and interactions have been irrevocably changed. The scale and speed of these changes is such that the gap between available knowledge and the planning of restoration measures is widening. As such, restoration is no more than a mitigation process or the alleviation of symptoms. The experience gathered to date suggests that the pollution-induced decline of forests can be slowed down and perhaps even reversed at most stages of degradation. In the case of continuing pollution, its effects can be mitigated to prevent further damage, but the initiation of the recovery process requires a decrease in the pollution load.

In forest restoration, small-scale, short-term experiments dominate over practical applications. Thus it is reasonably well known how to initiate the early stages of rehabilitation, but there are substantial uncertainties over the longer-term effects. The latter especially concern possible side-effects and ecological risks which may result from human intervention, such as increased leaching of toxic substances and/or nutrients. To reduce these risks, a careful analysis of the need for chemical treatments such as liming or fertilizing should be carried out for each specified area.

Restoration of forests after the reduction of emissions is based on the suggestion that some successional stages are not essential and can be bypassed by proper intervention (Ashby, 1987). Thus, in practical terms, the rehabilitation of forests damaged by pollution is an attempt to partially substitute time with money. However, in view of the possible adverse effects, a suitable strategy should be developed for each site in question. We suggest that whenever possible a 'minimal intervention' approach (Skaller, 1981) aimed at promoting natural succession should be used instead of an artificial re-creation of the desired ecosystem.

Acknowledgements

We are grateful to H.-S. Helmisaari, J.L. Innes, N.V. Lukina, E. Mälkönen, I.A. Paraketzov, A.V. Selikhovkin, and V.T. Yarmishko for providing us with valuable information, and an anonymous referee for the constructive criticism of an earlier draft of the Chapter. The writing of the manuscript was supported by the Academy of Finland (RESTORE research programme).

References

Agerer, R. (1991) Streuzersetzende Großpilze im Högwaldprojekt: Reaktionen im vierten Jahr der Behandlung. *Forstwissenschaftliche Forschungen* 39, 99–102.

Ahrens, M. (1994) The liverwort *Anastrophyllum michauxii* (F. Web.) Buch in the Black Forest (Southwestern Germany). *Herzogia* 10, 115–119.

Aldinger, E. (1987) Elementgehalte im Boden und in Nadeln verschieden stark geschädigter Fichten-Tannen-Bestände auf Praxiskalkungsflächen im Buntsandstein-Schwarzwald. *Freiburger Bodenkundliche Abhandlungen* 19, 1–266.

Alexeyev, V.A., Bogdan, M.A., Kataev, O.A., Popovichev, B.G., Selikhovkin, A.V. and Shenderova, S.S. (1985) Investigation of the forest state and forest regeneration in areas of human impact in Bratsk and Ust-Ilimsk forest economical units and elaboration of measures of forest protection. Unpublished report, State Forest Technical Academy, Leningrad (in Russian).

Andersson, F. and Persson, T. (1988) *Liming as a Measure to Improve Soil and Tree Condition in Areas Affected by Air Pollution – Results of an Ongoing Research Programme.* Report 3518, National Swedish Environmental Protection Board, Stockholm.

Ashby, W.C. (1987) Forests. In: Jordan, W.R., Gilpin, M.E. and Aber, J.D. (eds) *Restoration Ecology: a Synthetic Approach to Ecological Research.* Cambridge University Press, Cambridge, pp. 89–108.

Baule, H. and Fricker, C. (1967) *Die Düngung von Waldbäumen.* BLV Bayerischer Landwirtschaftsverlag GmbH, München.

Bergseth, M. (1978) Verteilung von Gesamt-Schwefel und Sulfationen verschiedener Bindungsstärke in Norwegischen Waldböden. *Acta Agricoltura Scandinavica* 28, 313–322.

Blaylock, M.J., Salt, D.E., Dushenkov, S., Zakharova, O., Gussman, C., Kapulnik, Y., Ensley, B.D. and Raskin, I. (1997) Enhanced accumulation of Pb in Indian mustard by soil-applied chelating agents. *Environmental Science and Technology* 31, 860–865.

Bradshaw, A.D. (1997) Restoration of mined lands – using natural processes. *Ecological Engineering* 8, 255–269.

Bridges, E.M. (1991) Reclaiming contaminated land in the city of Swansea. *Agricultural Engineering* 46, 115–117.

Bryant, J.P., Chapin, F.S. III and Klein, D.R. (1983) Carbon/nutrient balance of boreal plants in relation to vertebrate herbivory. *Oikos* 40, 357–368.

Castrén, M., Haavisto, T., Leinonen, K.P., Poutanen, H. and Nuorteva, P. (1990) The influence of soil preparation on the concentrations of iron, aluminium, zinc and cadmium in leaves of *Betula* spp. and *Maianthemum bifolium*. *Annales Botanici Fennici* 27, 139–143.

Crooke, W.M. and Inkson, R.H.E. (1955) The relationship between nickel toxicity and major nutrient supply. *Plant and Soil* 6, 1–15.

Cunningham, S.D., Berti, W.R. and Huang, J.W. (1995) Phytoremediation of contaminated soils. *Trends in Biotechnology* 13, 393–397.

DeGoede, R.G.M. and Dekker, H.H. (1993) Effects of liming and fertilization on nematode communities in coniferous forest soils. *Pedobiologia* 37, 193–209.

Derome, J.R.M. and Lindroos, A.-J. (1998) Effects of heavy metal contamination on macronutrient availability and acidification parameters in forest soil in the vicinity of the Harjavalta Cu–Ni smelter, SW Finland. *Environmental Pollution* 99, 225–232.

Derome, J., Kukkola, M. and Mälkönen, E. (1986) *Forest Liming on Mineral Soils. Results of Finnish Experiments*. Report 3084, National Swedish Environmental Protection Board, Stockholm.

Downing, R.J., Hettelingh, J.-P. and de Smet, P.A.M. (eds) (1993) *Calculation and Mapping of Critical Loads in Europe: Status Report 1993*. Coordination Center for Effects, RIVM Report 259101003, National Institute of Public Health and Environmental Protection, Bilthoven.

Eltrop, L., Brown, G., Joachim, O. and Brinkmann, K. (1991) Lead tolerance of *Betula* and *Salix* in the mining area of Mechernich/Germany. *Plants and Soil* 131, 275–285.

Erich, M.S. and Trusty, G.M. (1997) Chemical characterization of dissolved organic matter released by limed and unlimed forest soil horizons. *Canadian Journal of Soil Science* 77, 405–413.

Evdokimova, G.A. (1995) *Ecological – Microbiological Foundations of Soil Protection in the Far North*. Kola Science Centre, Apatity (in Russian).

Evers, F.H. and Hüttl, R.F. (1990) A new fertilization strategy in declining forests. *Water, Air, and Soil Pollution* 54, 495–508.

Fanta, J. (1997) Rehabilitating degraded forests in Central Europe into self-sustaining forest ecosystems. *Ecological Engineering* 8, 289–297.

Fanta, J. and Kobus, A. (1995) Restoration of forest ecosystems in Krkonose mountains: preliminary results of an inventory study. In: Tesar, V. (ed.) *Management of Forests Damaged by Air Pollution. Proceedings of the Workshop P2.05–07, Silviculture in Polluted Areas Working Party, Trutnov, Czech Republic, 5–9 June 1994*. University of Agriculture, Brno, pp. 97–102.

Fiedler, H.J., Nebe, W. and Hoffmann, F. (1973) *Forstliche Pflanzenernährung und Düngung*. G. Fischer Verlag, Jena.

Först, K. and Kreutzer, K. (1977) Die neue regionale Standortsgliederung und ihre Bedeutung für Forstbetrieb und Landesplanung. *Forstwissenschaftliches Centralblatt* 96, 49–55.

Fritze, H., Vanhala, P., Pietikäinen, J. and Mälkönen, E. (1996) Vitality fertilization of Scots pine stands growing along a gradient of heavy metal pollution: short-term effect on microbial biomass and respiration rate of the humus layer. *Fresenius Journal of Analytical Chemistry* 354, 750–755.

Golutvina, L.S., Pavlov, I.N. and Polistchuk, O.A. (1987) Elaboration of the methods of forest restoration and improvement of forest resistance to damaging agents in surroundings of Bratsk Forestry Enterprise. Unpublished report, State Forest Technical Academy, Leningrad (in Russian).

Golutvina, L.S., Anikin, A.S., Pavlov, I.N. and Polistchuk, O.A. (1988) Elaboration of the methods of forest restoration and improvement of forest resistance to damaging agents in surroundings of Bratsk Forestry Enterprise. Unpublished report, State Forest Technical Academy, Leningrad (in Russian).

Gubalba, C.P. and Driscoll, C.T. (1991) Watershed liming as a strategy to mitigate acidic deposition in the Adirondack region of New York. In: Olem, H., Schreiber, R.H., Brocksen, R.W. and Porcella, D.B. (eds) *International Lake and Watershed Liming Practices*. Terrene Institute, Washington, DC, pp. 145–159.

Gunn, J.M. (ed.) (1995) *Restoration and Recovery of an Industrial Region: Progress in Restoring the Smelter-Damaged Landscape Near Sudbury, Canada*. Springer Verlag, New York.

Haukioja, E., Ossipov, V., Koricheva, J., Honkanen, T., Larsson, S. and Lempa, K. (1998) Biosynthetic origin of carbon-based secondary compounds: cause of variable responses of woody plants to fertilization? *Chemoecology* 8, 133–139.

Hausser, K. (1958) Waldbauliche und betriebswirtschaftliche Erfolge der Forstdüngung-erlaeutert an Beispielen aus dem Buntsandsteingebiet des württembergischen Schwarzwaldes. *Allgemeine Forstzeitschrift* 13, 125–130.

Helmisaari, H.-S., Derome, J., Fritze, H., Nieminen, T., Palmgren, K., Salemaa, M. and Vanha-Majamaa, I. (1995) Copper in Scots pine forests around a heavy-metal smelter in south-western Finland. *Water, Air, and Soil Pollution* 85, 1727–1732.

Helmisaari, H.-S., Makkonen, K., Olsson, M., Viksna, A. and Mälkönen, E. (1999) Fine-root growth, mortality and heavy metal concentrations in limed and fertilized *Pinus sylvestris* (L.) stands in the vicinity of a Cu–Ni smelter in SW Finland. *Plant and Soil* (in press).

Herms, D.A. and Mattson, W.J. (1992) The dilemma of plants: to grow or defend. *Quarterly Reviews in Biology* 67, 283–335.

Hildebrand, E.E. (1990) Der Einfluss von Forstdüngungen auf die Lösungsfracht des Makroporenwassers. *Allgemeine Forstzeitschrift* 45, 604–607.

Howells, G. and Dalziel, T.R.K. (eds) (1992) *Restoring Acid Water: Loch Fleet 1984–1990*. Elsevier Applied Sciences, London.

Hüttl, R.F. (1988a) Forest decline and nutritional disturbances. In: Cole, D.W. and Gessel, S.P. (eds) *Forest Site Evaluation and Long-Term Productivity*. University of Washington Press, Seattle, pp. 180–186.

Hüttl, R.F. (1988b) 'New type' forest decline and restabilization/revitalization strategies – a programmatic focus. *Water, Air, and Soil Pollution* 41, 95–111.

Hüttl, R.F. (1989) Liming and fertilization as mitigation tools in declining forest ecosystems. *Water, Air, and Soil Pollution* 44, 93–118.

Hüttl, R.F. (1991) Die Nährelementversorgung geschädigter Wälder in Europa und Nordamerika. *Freiburger Bodenkundliche Abhandlungen* 28, 1–440.

Hüttl, R.F. (1993a) Mg deficiency – a 'new' phenomenon in declining forests – symptoms and effects. In: Hüttl, R.F. and Mueller-Dombois, D. (eds) *Forest Decline in the Atlantic and Pacific Region*. Springer-Verlag, Berlin, pp. 97–114.

Hüttl, R.F. (1993b) Forest soil acidification. *Angewandte Botanik* 67, 66–75.

Hüttl, R.F. (1995) Gezielte Düngung in geschädigter Waldbeständen. *Forstliche Schriftenreihe, Universität Bodenkultur Wien* 9, 19–33.

Hüttl, R.F. and Fink, S. (1988) Diagnostische Düngungsversuche zur Revitalisierung geschädigter Fichtenbestände (*Picea abies* Karst.) in Südwestdeutschland. *Forstwissenschaftliches Centralblatt* 107, 173–183.

Hüttl, R.F. and Hunter, I. (1992) Nutrient management of forests under stress for improved health and increased productivity. *Fertilizer Research* 32, 71–82.

Hüttl, R.F. and Mehne, B.M. (1988) 'New type' of forest decline, nutrient deficiencies and the 'virus'-hypothesis. In: Mathy, P. (ed.) *Air Pollution and Ecosystems*. Reidel, Dordrecht, pp. 870–873.

Hüttl, R.F. and Zöttl, H.W. (1993) Liming as a mitigation tool in Germany's declining forests – reviewing results from former and recent trials. *Forest Ecology and Management* 61, 325–338.

Insam, H., Haselwandter, K., Berreck, M., Girschick, B. and Zechmeister-Boltenstern, S. (1994) Revitalisierungsdüngung von Fichtenbeständen: Einfluss auf Mikrobielle Umsetzungsprozesse und Biomasse. *Forstliche Schriftenreihe, Universität Bodenkultur Wien* 7, 83–97.

Kalatskaya, M.N. (1997) Nutritional regime of podzol soils in spruce forests of the Kola peninsula and its regulation under the impact of aerial emissions. Abstract of PhD thesis, Moscow State University, Moscow (in Russian).

Katzensteiner, K., Glatzel, G., Kazda, M. and Sterba, H. (1992) Effects of air pollutants on mineral nutrition of Norway spruce and revitalization of declining stands in Austria. *Water, Air, and Soil Pollution* 61, 309–322.

Katzensteiner, K., Eckmüllner, O., Jandl, R., Glatzel, G., Sterba, H., Wessely, A. and Hüttl, R.F. (1994) Revitalisierungsdüngung von Fichtenbeständen: Einfluss auf Bodenwasser und Baumernährung. *Forstliche Schriftenreihe, Universität Boden-kultur Wien* 7, 67–82.

Katzensteiner, K., Eckmüllner, O., Jandl, R., Glatzel, G., Sterba, H., Wessely, A. and Hüttl, R.F. (1995) Revitalization experiments in magnesium deficient Norway spruce stands in Austria. *Plant and Soil* 168/169, 489–500.

Kaupenjohann, M. (1991) Magnesium-Mangel im Wald. *Kali-Briefe* (Büntehof) 20, 561–577.

Kaupenjohann, M. and Zech, W. (1989) Waldschäden und Düngung. *Allgemeine Forstzeitung* 37, 1002–1008.

Klimo, E.D.V. (1991) Acidification and liming of forest soils in the Beskids (Czechoslova-kia). *Lesnictvi* (Prague) 37, 61–72 (in Czech).

Kozlov, M.V. (1997) Pollution impact on insect biodiversity in boreal forests: evaluation of effects and perspectives of recovery. In: Crawford, R.M.M. (ed.) *Disturbance and Recovery in Arctic Lands: An Ecological Perspective. Proceedings of the NATO Advanced Research Workshop on Disturbance and Recovery of Arctic Terrestrial Ecosystems, Rovaniemi, Finland, 24–30 September 1995.* NATO ASI Series, Partnership subseries 2, Environment, Vol. 25. Kluwer Academic Publishers, Dordrecht, pp. 213–250.

Kozlov, M.V. and Haukioja, E. (1999) Performance of birch seedlings replanted in heavily polluted industrial barrens of the Kola peninsula, NW Russia. *Restoration Ecology* (in press).

Kozlov, M.V., Zvereva, E.L. and Niemelä, P. (1999) Effects of soil quality and air pollution on rooting and survival of *Salix borealis* cuttings. *Boreal Environment Research* 4, 67–76.

Kreutzer, K. (1986) Zusammenfassende Diskussion der Ergebnisse aus experimentellen Freiland-Untersuchungen über den Einfluß von sauren Niederschlägen und Kalkung in Fichtenbeständen (*Picea abies* [L.] Karst.). *Forstwissenschaftliches Centralblatt* 105, 371–379.

Kreutzer, K. (1995) Effects of forest liming on soil processes. *Plant and Soil* 168/169, 447–470.

Kreutzer, K. and Schierl, R. (1992) Versuche mit dolomitischem Kalk im Höglwald. In: Glatzel, G., Tandl, R. and Hager, H. (eds) *Magnesiummangel im Mitteleuropäischen Waldökosystemen. Forstliche Schriftenreihe Universität Bodenkultur Wien* 5, 171–186.

Kreutzer, K., Schierl, R., Reiter, H. and Göttlein, A. (1989) Effects of irrigation and compensatory liming in a Norway spruce stand (*Picea abies* [L.] Karst.). *Water, Air, and Soil Pollution* 48, 111–125.

Kubelka, L., Karásek, A., Rybár, V., Badalík, V. and Slodicák, M. (1993) *Forest Regeneration in the Heavily Polluted NE 'Krusné Hory' Mountains.* Czech Ministry of Agriculture, Prague.

Kukkola, M. and Saramäki, J. (1983) Growth response in repeatedly fertilized pine and spruce stands on mineral soils. *Communicationes Instituti Forestalis Fenniae* 114, 1–55.

Kytö, M., Niemelä, P. and Annila, E. (1998) Effects of vitality fertilization on the resin flow and vigour of Scots pine in Finland. *Forest Ecology and Management* 102, 121–130.

Latocha, E. (1990) Scots pine stands reconstruction possibilities in the Upper Silesia industrial region. In: *XIX World Congress of IUFRO, Montréal, Canada, 5–11 August 1990. Proceedings, Division 2.* Oxford Forestry Institute, Oxford, pp. 447–454.

Lautenbach, W.E., Miller, J., Beckett, P.J., Negusanti, J.J. and Winterhalder, K. (1995) Municipal land restoration program: the regreening process. In: Gunn, J.M. (ed.) *Restoration and Recovery of an Industrial Region: Progress in Restoring the Smelter-Damaged Landscape Near Sudbury, Canada.* Springer Verlag, New York, pp. 109–122.

Lee, D. and Woo, S.-Y. (1998) Status of waste generation, disposal ways and poplar species as a landfill cover in South Korea. In: Sassa, K. (ed.) *Environmental Forest Science.* Kluwer, Dordrecht, pp. 83–92.

Lenz, R. and Haber, W. (1992) Approaches for the restoration of forest ecosystems in northeastern Bavaria. *Ecological Modelling* 63, 299–317.

Liu, J.C. and Ende, H.P. (1989) Wirkungen von Nitrat- und Sulfatdüngern auf Boden-pH und Nährelementversorgung von Fichten- und Buchenbeständen. *Kernforschungszentrum Karlsruhe, KfK-PEF Bericht* 55, 323–336.

Liu, J.C. and Trüby, P. (1989) Bodenanalitische Diagnose von K- und Mg-Mangel in Fichtenbestanden (*Picea abies* Karst.). *Zeitschrift für Pflanzenernährung und Bodenkunde* 152, 307–311.

Lorio, P.L. (1986) Growth-differentiation balance: a basis for understanding southern pine beetle-tree interactions. *Forest Ecology and Management* 14, 259–273.

Louda, S.M. (1995) Insect pests and plant stress as considerations for revegetation of disturbed ecosystems. In: Cairns, J. (ed.) *Rehabilitating Damaged Ecosystems.* Lewis Publishers, Boca Raton, Florida, pp. 335–356.

Luganskij, N.A., Terekhov, G.G., Sokolova, N.M., Menstchikov, S.L. and Srodnykh, T.B. (1983) Investigation of suitability of soils polluted by magnesite dusts for forest restoration, and recultivation of polluted areas. Unpublished report, Ural Forest Experimental Station, All-Union Research Institute of Forestry and Forest Mechanization, Sverdlovsk (in Russian).

Lukina, N.V. and Nikonov, V.V. (1998) Nutrient status of the northern forests and approaches to revitalization of lands subjected to air pollution. In: *Assessing the Consequences of Global Changes for the Barents Region: the Barents Sea Impact Study (BASIS). First International Research Conference, St Petersburg, AARI, 22–25 February 1998. Terrestrial Environment, Ecology and Forestry. Abstracts.* Arctic and Antarctic Research Institute, St Petersburg. p. 12.

Lukina, N., Nikonov, V.V. and Belova, E. (1997) A pilot study on the rehabilitation of lands subjected to air pollution in the Kola peninsula. Unpublished report, Institute of North Industrial Ecology Problems, Apatity.

Makeschin, F. (1991) Auswirkungen von saurer Beregnung und Kalkung auf die Regenwurmfauna (Lumbricidae: Oligochaeta) im Fichtenwaldbestand Högwald. *Forstwissenschaftliche Forschungen* 39, 117–128.

Mälkönen, E., Derome, J., Fritze, H., Helmisaari, H.-S., Kukkola, M., Kytö, M., Saarsalmi, A. and Salemaa, M. (1999) Compensatory fertilization of Scots pine stands polluted by heavy metals. *Nutrient Cycling in Agroecosystems* (in press).

Marschner, B. (1990) Elementumsätze in einem Kiefernforstökosystem auf Rostbraunerde unter dem Einflußeiner Kalkung/Düngung. *Berichte der Forschungszentrum Waldökosysteme* A60, 1–192.

Marschner, B. (1993) Microbial contribution to sulphate mobilization after liming an acid forest soil. *Journal of Soil Science* 44, 459–466.

Marschner, B., Stahr, K. and Renger, M. (1989) Potential hazards of lime application in a damaged pine forest ecosystem in Berlin, Germany. *Water, Air, and Soil Pollution* 48, 45–57.

Marschner, B., Stahr, K. and Renger, M. (1992) Lime effects on pine forest floor leachate chemistry and element fluxes. *Journal of Environmental Quality* 21, 410–419.

Matyssek, R. and Innes, J.L. (1999) Ozone – a risk factor for trees and forests in Europe? *Water, Air, and Soil Pollution* (in press).

Matzner, E. and Meiwes, K.J. (1990) Effects of liming and fertilization on soil solution chemistry in Northern German forest ecosystems. *Water, Air, and Soil Pollution* 54, 377–389.

Mitsch, W.J. and Mander, Ü. (1997) Remediation of ecosystems damaged by environmental contamination: applications of ecological engineering and ecosystem restoration in Central and Eastern Europe. *Ecological Engineering* 8, 247–254.

Mitscherlich, G. and Wittich, W. (1958) Düngungsversuche in älteren Beständen Badens. *Allgemeine Forst- und Jagdzeitung* 41, 1151–1154.

Moiseenko, T. (1994) Acidification and critical loads in surface waters: Kola, Northern Russia. *Ambio* 23, 418–424.

Newson, M. (1995) The Earth as output: pollution. In: Johnston, R.J., Taylor, P.J. and Watts, M.J. (eds) *Geographies of Global Change: Remapping the World in the Late Twentieth Century.* Blackwell, Oxford, pp. 333–353.

Nikonov, V.V., Lukina, N.V. and Kalatskaya, M.N. (1999) Correction of nutritional status of defoliating spruce forests. *Lesovedenie* [*Forestry; Moscow*] (in press) (in Russian).

Nuorteva, P. (1990) *Metal Distribution Patterns and Forest Decline. Seeking Achilles' Heels for Metals in Finnish Forest Biocoenoses.* Publications of the Department of Environmental Conservation, No. 11. University of Helsinki, Helsinki.

Nyström, U., Hultberg, H. and Lind, B.B. (1995) Can forest-soil liming mitigate acidification of surface waters in Sweden? *Water, Air, and Soil Pollution* 85, 1855–1860.

Päivinen, L. (1994) The experiments at Tammela, Sumiainen and Saarijärvi. In: *Effect of Fertilization on Forest Ecosystem.* Biological Research Report, University of Juväskylä, No. 38, pp. 173–182.

Papen, H., von Berg, R., Hellmann, B. and Rennenberg, H. (1991) Einfluß von saurer Beregnung und Kalkung auf chemolitotrophe und heterotrophe Nitrifikation in Böden des Högwaldes. *Forstwissenschaftliche Forschungen* 39, 111–116.

Pawlowski, L. (1997) Acidification: its impact on the environment and mitigation strategies. *Ecological Engineering* 8, 271–288.

Persson, T., Rudebeck, A. and Wiren, A. (1995) Pools and fluxes of carbon and nitrogen in 40-year-old forest liming experiments in southern Sweden. *Water, Air, and Soil Pollution* 85, 901–906.

Pietelka, L.F. (1988) Evolutionary responses of plants to anthropogenic pollutants. *Trends in Ecology and Evolution* 3, 233–236.

Potter, M.J. (1991) *Treeshelters.* Forestry Commission Handbook 7, HMSO, London.

Pysek, P. (1992) Dominant species exchange during succession in reclaimed habitats: a case study from areas deforested by air pollution. *Forest Ecology and Management* 54, 27–44.

Pysek, P. (1994) Effect of soil characteristics on succession in sites reclaimed after acid rain deforestation. *Ecological Engineering* 3, 39–47.

Rachwal, L. and Wit-Rzepka, M. (1989) Responses of birches to pollution from a copper smelter. Part II. Results of field experiment. *Arboretum Kórnickie* 34, 185–205.

Ranft, H. (1982) Düngung als Anpassungsmaßnahme in Imissionsschadgebiet Oberes Erzgebirge. *Beiträge der Forstwirtschaft* 16, 110–122.

Raskin, I. (1996) Plant genetic engineering may help with environmental cleanup. *Proceedings of the National Academy of Sciences, USA* 93, 3164–3166.

Schaaff, W. (1995) Effects of Mg(OH)$_2$-fertilization on nutrient cycling in a heavily damaged Norway spruce ecosystem (NE Bavaria/FRG). *Plant and Soil* 168/169, 505–511.

Schornick, O. (1989) Resultat einer langjährigen Sukzession der Bodenvegetation auf Düngeversuchsflächen in Fichtenforsten des Nordschwarzwaldes. In: *IMA-Quertschnittseminar 'Düngung geschädigter Wälder', KfK-PEF* Vol. 55, 87–96.

Schulze, E.D. (1989) Air pollution and forest decline in a spruce (*Picea abies*) forest. *Science* 244, 776–783.

Shumakov, V.S. (1977) Increase of forest productivity by application of mineral fertilizers. In: *Increase of Forest Productivity by Silvicultural Measures.* All-Union Research Institute for Silviculture and Mechanisation of Forestry, Moscow, pp. 45–58 (in Russian).

Skaller, P.M. (1981) Vegetation management by minimal interference: working with succession. *Landscape Planning* 8, 149–174.

Skelly, J.M., Innes, J.L., Savage, J.E., Snyder, K.R., VanderHeyden, D., Zhang, J. and Sanz, M.J. (1999) Observation and confirmation of foliar ozone symptoms on native plant species of Switzerland and southern Spain. *Water, Air, and Soil Pollution* (in press).

Skraba, D. (1989) Effects of surface liming of soils on stream flow chemistry in denuded acid, metals contaminated watershed near Sudbury, Ontario. MSc thesis, Laurentian University, Sudbury, Ontario (cited in Lautenbach *et al.*, 1995).

Slodicák, M. (1994) Thinning of air polluted stands of Norway spruce. In: Tesar, V. (ed.) *Management of Forests Damaged by Air Pollution. Proceedings of the Workshop P2.05–07, Silviculture in Polluted Areas Working Party, Trutnov, Czech Republic, 5–9 June 1994.* University of Agriculture, Brno, pp. 35–39.

Smolander, A., Kurka, A.-M., Kitunern, V. and Mälkönen, E. (1994) Microbial biomass C and N, and respiratory activity in soil of repeatedly limed and N- and P-fertilized Norway spruce stands. *Soil Biology and Biochemistry* 26, 957–962.

Smolyaninov, I.I. (1969) *Biological Rotation of Elements and Increase in Forest Productivity.* Lesnaya Promyshlennost, Moscow (in Russian).

Stchavrovskij, V.A., Zaparanjuk, A.E. and Sbitneva, L.D. (1983) Impact of toxic pollutants on growth and health status of suburban forests in Ural region, and revitalization of polluted stands by fertilization. Unpublished report, Ural Forest Technological Institute, Sverdlovsk (in Russian).

Stomp, A. M., Han, K. H., Wilbert, S. and Gordon, M.P. (1993) Genetic improvement of tree species for remediation of hazardous wastes. *In Vitro Cellular and Developmental Biology of Plants* 29P, 227–232.

Tesar, V. (1994) *Silvicultural Systems for Forests Impacted by Air Pollution in the Region of Trutnov.* University of Agriculture, Brno.

Tesar, V. and Tichy, J. (1990) Results and new objectives in restoring the forest damaged by air pollution in Bohemian mountains. In: *XIX World Congress of IUFRO, Montréal, Canada, 5–11 August 1990. Proceedings, Division 2.* Oxford Forestry Institute, Oxford, pp. 455–462.

Thomasius, H. (1990) Conversion and preservation of forests damaged by air pollution in the highlands of the GDR. In: *XIX World Congress of IUFRO, Montréal, Canada, 5–11 August 1990. Proceedings, Division 2.* Oxford Forestry Institute, Oxford, pp. 463–476.

Tsvetkov, V.F. (1987) To the problem of forest restoration in the zone of industrial emission impact in the Kola peninsula. In: Karpovich, V.N. (ed.) *The Problems of Investigation and Nature Protection in the White Sea Region.* Murmansk Publishing House, Murmansk, pp. 148–156 (in Russian).

Tsvetkov, V.F. and Chekrisov, E.A. (1987) An experience of forest recultivation in areas affected by industrial emissions in Kola Peninsula. In: Krivolutskij, D.A. (ed.) *Impact of Industrial Enterprises on the Environment.* Nauka, Moscow, pp. 112–119 (in Russian).

Tveite, B., Abrahamsen, G. and Stuanes, A.O. (1991) Liming and wet acid deposition effects on tree growth and nutrition: experimental results. In: Zöttl, H.W. and Hüttl, R.F. (eds) *Management of Nutrition in Forest Under Stress.* Kluwer Academic Publishers, Dordrecht, pp. 409–422.

Ulrich, B. (1981) Theoretische Betrachtung des Ionenkreislauf in Waldökosystemen. *Zeitschrift für Pflanzenernährung und Bodenkunde* 144, 647–659.

Ulrich, B., Mayer, R. and Khanna, P.K. (1979) Deposition von Luftverunreinigungen und ihre Auswirkungen in Waldökosystemen im Solling. *Schriften der Forstliche Fakultät der Univeristät Göttingen und der Niedersächsiscge Forstliche Versuchsanstalt* 58, 258–260.

Utriainen, M.A., Kärenlampi, L.V., Kärenlampi, S.O. and Schat, H. (1997) Differential tolerance to copper and zinc of micropropagated birches tested in hydroponics. *New Phytologist* 137, 543–549.

Valtanen, J. and Engberg, M. (1987) The results from Kainuu and Pohjanmaa of the ploughed area reforestation experiment begun during 1970–72. *Folia Forestalis* 686, 1–42.

Väre, H. (1988) Correlation between mycorrhizae and the growth of *Pinus sylvestris* in ploughed sites in northern Finland. *Karstenia* 28, 31–33.

Wenzel, B. and Ulrich, B. (1988) Kompensationskalkung – Riskien und ihre Minimierung. *Forst- und Holzwirtschaft* 41, 12–16.

Whitby, L.M., Stokes, P.M., Hutchinson, P.C. and Myslik, G. (1976) Ecological consequences of acidic and heavy-metal discharges from the Sudbury smelters. *Canadian Mineralogy* 14, 47–57.

Winterhalder, K. (1984) Environmental degradation and rehabilitation in the Sudbury area. *Laurentian University Reviews* 16, 15–47.

Winterhalder, K. (1988) Trigger factors initiating natural revegetation processes on barren, acid, metal-toxic soils near Sudbury, Ontario smelters. In: *Mine Drainage and Surface Mine Reclamation. Proceedings of a Conference Held in Pittsburgh, Pennsylvania 19–21 April 1988*, Vol. II, *Mine Reclamation, Abandoned Mine Lands and Policy Issues*, pp. 118–124.

Winterhalder, K. (1995) Natural recovery of vascular plant communities on the industrial barrens of the Sudbury area. In: Gunn, J.M. (ed.) *Restoration and Recovery of an Industrial Region: Progress in Restoring the Smelter-Damaged Landscape Near Sudbury, Canada*. Springer Verlag, New York, pp. 93–102.

Yan, N.D., Keller, W. and Gunn, J.M. (1995) Liming of Sudbury lakes: lessons for recovery of aquatic biota from acidification. In: Gunn, J.M. (ed.) *Restoration and Recovery of an Industrial Region: Progress in Restoring the Smelter-Damaged Landscape Near Sudbury, Canada*. Springer Verlag, New York, pp. 195–204.

Zelles, L., Stepper, K. and Zsolnay, A. (1990) The effect of lime on microbial activity in spruce (*Picea abies* L.) forests. *Biology and Fertility of Soils* 9, 78–82.

Zöttl, H.W. (1990) Ernährung und Düngung der Fichte. *Forstwissenschaftliches Centralblatt* 106, 105–114.

Zöttl, H.W. and Hüttl, R.F. (1986) Nutrient supply and forest decline in southwest-Germany. *Water, Air, and Soil Pollution* 31, 449–462.

Zöttl, H.W. and Larichev, O.I. (eds) (1991) *Management of Nutrition in Forest Under Stress*. Kluwer Academic Publishers, Dordrecht.

Zvereva, E.L., Kozlov, M.V. and Haukioja, E. (1997) Population dynamics of a herbivore in an industrially modified landscape: case study with *Melasoma lapponica* (Coleoptera: Chrysomelidae). *Acta Phytopatologia Entomologia Hungariensis* 32, 251–258.

International Activities to Reduce Pollution Impacts at the Regional Scale

12

K. Bull[1] and G. Fenech[2]

[1]ITE Monks Wood, Huntingdon, UK; [2]Environment Canada, Downsview, Ontario, Canada

At the regional scale there has been much effort made to introduce international agreements on pollution control over the last two decades. In Europe in the 1980s, much of the early focus was upon simple flat rate controls based upon emission limits or technology. Some countries adopted the best available technology or the precautionary principle for their emission controls. In the 1990s, the move towards effects-based approaches have made use of the development of critical loads maps and our knowledge of long-range atmospheric transport resulting from models and measurements at monitoring sites. An effects-based sulphur protocol for the UNECE Convention on Long-Range Transboundary Air Pollution is now being followed by a multi-pollutant, multi-effect protocol which will address effects of sulphur, nitrogen and ozone. This will be supported by European Community strategies on acidification and ozone effects (which will also consider the effects of excess nutrient nitrogen) and emission ceilings directives. In North America bilateral agreements have been used to reduce sulphur emissions and their impacts internationally; Canada has indicated that further action will be needed in the future to address outstanding acidification problems. In general, emissions have fallen in line with protocol obligations across Europe, while similar changes have taken place in North America. There are differences for many individual countries, however, with some achieving more while others, usually those who are not parties to particular protocols, much less. Overall, sulphur emissions will probably be around 20% of their maximum in the mid-1980s by 2010; this has benefits both internationally but also more locally where individual sources often have significant impacts on

ecosystems. Improvements in sulphur emissions are placing more emphasis on the need to control nitrogen pollutant emissions. In addition, future
consideration will need to be taken of the large costs associated with high
levels of protection and whether that protection is justifiable or attainable;
for this, better information on benefits will be needed.

1 Introduction

At the Stockholm Conference in 1972 it was fully acknowledged, perhaps
for the first time, that international agreements would be important for the
control of the effects of atmospheric pollution. Evidence showed that many
atmospheric pollutants could travel extremely long distances, and as a result
they might contribute to problems such as acidification and nutrient nitrogen
effects in countries far from those in which they were emitted. In the following
years much thought was given to suitable mechanisms by which environmental effects resulting from atmospheric pollutants might be avoided.

While at a local level there may be opportunities to reduce the impacts of
pollution on the natural and managed environment, these often do not address
the regional issues where long-range transport is involved. Taking account of
significant local sources of emissions may lead to planned reductions through
changes in technology or reduced time of operation. For example, where
atmospheric concentrations of pollutants are significant, increased stack
height may reduce ground concentrations of SO_2 to non-harmful levels and
local effects may be minimized. However, the long-range transport of emissions still remains unless more costly measures are introduced to decrease
emissions and, for these, international agreement and legislation have played
a major role.

This chapter considers three areas of international agreement which have
been significant for the northern hemisphere. The first is that initiated at the
Stockholm Conference and which resulted subsequently in the UNECE region
Convention on Long-Range Transboundary Air Pollution (LRTAP). The
Convention was signed by 34 governments in 1979 (UNECE, 1996) following
a high-level ministerial meeting on the protection of the environment. While
there was general agreement on the need to reduce long-range pollution
transport, the Scandinavian governments were particularly concerned about
acidification of soils and freshwaters, and in central Europe governments were
anxious to address a perceived problem of deteriorating forest health.

The other areas of international agreement dealt with here are not
described in great detail. These agreements have taken place in parallel with
discussions within the LRTAP Convention by countries who are parties to the
Convention. In North America, Canada and the USA have turned to bilateral
agreements to address transboundary pollution issues. They fall outside the

region where air pollution is modelled for the LRTAP Convention and, while remaining interested in the discussions on pollution control and becoming signatories of several protocols to the Convention itself, their cuts in emissions have not been linked directly to the discussions focused on Europe. The third area, the European Community (EC), also a signatory of the LRTAP Convention, has developed international strategies and legislation for its member states to reduce emissions from polluting sources. At times this has been done independently of the discussions for the UNECE region, but increasingly there has been a realization that a more coordinated approach has benefits for member states.

2   Flat rate and technology-driven policies

The Convention on Long-Range Transboundary Air Pollution entered into force in 1983 and was soon extended by a protocol on 'the reduction of sulphur emissions and their transboundary fluxes by at least 30%'. Sulphur emissions and their acidifying effects were seen as the highest priority for control measures. At that time, sulphur emissions had been progressively increasing in Europe and had continued to increase quite significantly prior to the period of protocol negotiations (Fig. 12.1). Peak emissions around 1980 coincided with the time when many governments were agreeing (i.e. ratifying) the conditions of the Convention and starting to consider possible action on a protocol for sulphur.

While there was disagreement in some countries, notably the UK, with regard to the extent and significance of long-range transport of air pollution, most governments accepted that pollution controls would be needed in the coming years. By 1993 (the target date for the sulphur protocol) most countries in Europe and North America, including the UK which did not sign the protocol, had achieved at least a 30% reduction in emissions from the levels in 1980. In many cases this was a much greater reduction from the maximum emissions which occurred in the mid-1980s. Nevertheless there were several European countries, all non-parties to the protocol, who did not reach the 30% objective.

The protocol itself provided no guidance on the means by which emissions should be reduced. Different countries had their own ideas on methods of reduction. In some, *best available technology* (BAT) was seen as the way of ensuring that emissions were minimized and environmental benefits were gained. This was seen as applying the *precautionary principle*; ensuring that damage was minimized by taking all steps to reduce emissions. Indeed, for some countries, such as Switzerland and Germany, this was the overriding driving force for reducing emissions, not their obligations to international agreements. In other countries, the costs of introducing control measures were considered in relation to their international obligations, while other

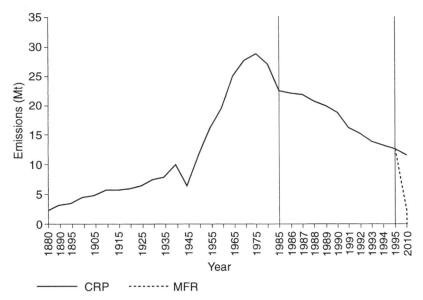

Fig. 12.1.   Past, present and future trends in sulphur emissions in Europe. Early data are simple estimates. Future emissions are shown in relation to current reduction plans (CRP) and what are believed to be maximum feasible reductions (MFR). (Data prepared by M. Posch from data provided by EMEP, the LRTAP Convention's Cooperative Programme for Monitoring and Evaluation of the Long-Range Transmission of Air Pollutants in Europe.)

approaches included consideration of alternative energy sources, such as nuclear power.

In parallel with the sulphur protocol, the EC set about defining regulations for the emitting sources in its region. Most of the focus of regulation in the EC region has been to protect human health. Directives aim at setting limit values, for example for SO_2 and NO_2; the restriction of peak levels does not necessarily imply an overall reduction in emissions. However, it was recognized that large proportions of emissions of sulphur and nitrogen oxides were from well-defined sources such as the large industrial sources, e.g. power generation plants and oil refineries, and mobile sources such as motor vehicles. So specific directives have been devised to control fuel quality and releases from these sources, such as the Large Combustion Plant Directive.

In North America, SO_2 emissions peaked in the early 1970s. Early concern about air quality led to local emission controls (e.g. for the Sudbury smelter in Ontario, see Chapter 7, this volume), but concern about acidic deposition did not arise until the late 1970s when Europe was also beginning to consider the need for action. Through the 1980s Canadian policy was tied to progress on international agreements on emission abatement. By the mid-1980s the

Canadian federal and provincial governments finally took a unilateral course and assembled an acid rain programme. Formal federal–provincial agreements, signed in 1987 and 1988, established specific targets and timetables for reducing SO_2 emissions from the seven provinces east of Saskatchewan by about 50% from the 1980 baseline by 1994. Furthermore, Canada's Green Plan (1990) aimed to extend eastern Canadian controls beyond 1994 and to conclude a transboundary Air Quality Agreement with the United States. These initiatives were aimed at reducing wet sulphate deposition from levels as high as 40 kg ha^{-1} year^{-1} to no more than 20 kg ha^{-1} year^{-1}. This 'target load' was based on the best information available at that time, and was expected to protect aquatic environments that were moderately sensitive. On that basis Canada signed and ratified the 1995 sulphur protocol in 1996. Formal negotiations on a Canada–USA Air Quality Agreement began in the summer of 1990. While an accord was accomplished, the USA did not sign up to the LRTAP sulphur protocol and the agreement between Canada and the USA to negotiate did not initially produce serious negotiations. Eventually President Bush submitted to Congress a set of proposals to amend the Clean Air Act. These included the provision that emissions of SO_2 from utilities be reduced by about 50%. Formal Canadian–American bilateral talks began before the bill was signed into law. The Canadian–USA Air Quality Agreement, signed in March 1991, contained a mutual exchange of commitments largely based upon existing domestic policies or legislation to reduce emissions of SO_2.

The success of the sulphur protocol and associated steps taken in countries throughout Europe and North America has been seen as a major success story for international pollution control. Although there are widely differing views on the role played by the Convention, and whether the Convention or individual counties were taking the leading role in taking steps to cut emissions, the fact remains that there was major international agreement and, in general, emissions have fallen in line with the targets set by the protocol. This has had benefits, not just with long-range transboundary pollution, but also locally where management of individual sources led to direct improvements of the pollution stress on the environment. It is true that decreases in emissions were not uniform across Europe and North America. Some countries, which adopted best available technology and the precautionary principle, achieved much more than was required. Other countries, mostly those that were non-parties to the protocol, failed to achieve a 30% decrease (ECE, 1995). However, the general trend in decreased emissions of sulphur has had significant benefits for most areas of Europe.

Following the steps taken to control sulphur emissions, it was recognized that oxides of nitrogen were likely to contribute increasingly to the acidification of the environment. So, in 1988, another 'flat rate' protocol was added to the LRTAP Convention concerning 'the control of emissions of nitrogen oxides or their transboundary fluxes'. The basic obligations of this protocol were much more specific and identified the need to apply national emission standards to stationary and mobile sources of nitrogen oxides. However, the ambition level was somewhat lower. The obligation was for limiting emissions by 1996 to those of 1987, in other words ensuring a levelling off of the increases of nitrogen oxide emissions which were experienced in earlier years. This was in recognition of the known problems for substantially reducing nitrogen oxide emissions in the near future. The protocol also differed from the sulphur protocol by providing a technical annex specifying the methods that might be used to meet the obligations under the protocol. Twenty-five parties to the Convention signed up to the protocol and 25 parties had ratified or accepted the protocol by 1994; for North America, Canada ratified and the USA accepted the protocol in 1991. Again many parties have met their obligations, but some countries who are parties to the protocol, and several of those who are not, failed to achieve the targets set (ECE, 1995). Overall, however, Europe and North America did achieve their goals.

A third flat-rate protocol to the LRTAP Convention was introduced in 1991 for the control of emissions of volatile organic compounds (VOCs) which, as the protocol notes, contribute with nitrogen oxides to the formation of tropospheric ozone, and all three pollutants can affect air quality as a result of transportation across international boundaries. The obligations amount to a 30% reduction in emissions from 1988 to 1999, although provision for the levelling off of emissions was made for some areas (e.g. where countries' emissions are small). While seeking to reduce the concentrations of VOCs in the air by limiting their emissions, there was no attempt to link this directly with the formation of tropospheric ozone or take account of any possible environmental impacts from this gas.

These early attempts at controlling the major air pollutants were supported by national legislation and by legislation in the EC where several countries were bound by legal instruments to take specific actions with regard to the quantity of their emissions or the type of processes allowed. However, none of these took environmental effects specifically into account, and there was no attempt to optimize environmental benefits from any action taken.

3   Effects-based approaches

The 1988 nitrogen oxide protocol included, as a basic obligation, the cooperation of parties to establish critical loads and identified for future work a 'critical loads' approach for determining appropriate control strategies. Similarly, the 1991 VOC protocol contains the obligation to cooperate in order to establish 'critical levels for photochemical oxidants' for future control strategies for VOCs. In this way it was recognized that future steps towards the control of long-range transport of pollution should be effects based and make use of the 'critical loads/levels' concept which had been discussed for some years prior to these protocols. At a bilateral level, the Canadians had put forward the notion of specifying critical loads in their accord with the USA, but this was firmly rejected by the American negotiators. Such loads were, they said, much too speculative at the present level of scientific understanding about ecosystem effects.

Critical loads were probably mentioned at the Stockholm conference in 1972 but there is no record of this. It is in the 1990s that ideas really developed regarding the critical amounts, loads (deposition) and levels (air concentrations), which might be considered the thresholds for effects and which might be seen as the ultimate targets for pollution controls. The basic ideas and the way they have developed have been discussed elsewhere (e.g. Bull, 1995), but the outcome in the LRTAP Convention was to introduce an effects-based critical loads approach into the revision of the sulphur protocol. The second sulphur protocol was signed in Oslo in 1994 and has recently been ratified by the necessary 16 parties to bring it into force; for North America this includes Canada but not the USA.

The second sulphur protocol is very different from earlier protocols. It is based upon studies of environmental information, the critical loads for soils and waters across Europe, and the use of atmospheric transport patterns and costs of abatement to identify least cost, maximum benefit scenarios. The various components are brought together in integrated assessment models (Hordijk, 1995; Hordijk and Kroeze, 1997) which are used to optimize spending to achieve benefits in a cost-effective way. Various scenarios were considered but a final agreed scenario, which aimed at reducing sulphur deposition towards the lowest critical loads in Europe, defined the obligations of countries in terms of the necessary sulphur emissions reductions required to meet the targets set. In practice, the lowest critical loads were defined by the 5-percentile in each modelling grid area (i.e. that critical load which would protect 95% of the ecosystems) and the reduction towards these values was set at 60% from the deposition resulting from 1990 emission levels of sulphur.

The EC has been keen to address the issues identified in the sulphur protocol through an effects-based approach featuring critical loads. To this end its Commission has set about defining an Acidification Strategy which aims to set targets analogous to, and at least as ambitious as, those identified in the LRTAP Convention protocols. For this it has used one of the same integrated assessment models used for the LRTAP deliberations and has drawn upon the same databases. As an ultimate goal the EC has agreed to seek to achieve critical loads across Europe, but intermediate targets based on ecosystem protection are being used to define national emissions ceilings. Recently, the EC has been keen to incorporate nutrient nitrogen effects (eutrophication) resulting from long-range transport of air pollution within the scope of the Acidification Strategy. While this may not be considered as a separate binding force within the integrated assessment modelling activities, the implications for eutrophication resulting from acidification model results will be an important element in the discussions on emission ceilings within the Community.

Directives on emission ceilings will be used to implement the Acidification Strategy, although final decisions on these will not be taken until further progress has been made towards a related Ozone Strategy (see below).

Even as the sulphur protocol was being negotiated it was recognized that acidification was not solely due to sulphur emissions. Both oxidized and reduced nitrogen deposition can also contribute to the acidification of soils and waters. In addition, it was appreciated that nitrogen deposition could result in excess nutrient effects to certain sensitive ecosystems, while nitrogen oxides and VOCs together can cause elevated tropospheric ozone levels (Grennfelt *et al.*, 1994). Such links between several pollutants and their effects have led to the development of what is known as the multi-pollutant, multi-effect protocol by parties under the LRTAP Convention. The protocol is seen as an extension to the 1988 nitrogen oxide protocol, but agreement has been reached to include both sulphur and VOCs, as well as reduced nitrogen emissions (ammonia), in the deliberations.

The EC's Ozone Strategy has developed in parallel with the discussions on the LRTAP Convention's multi-pollutant, multi-effect protocol described above. The European Commission, which is responsible for drawing up the strategy, is currently completing its work and will then consider the necessary legislation, the Directives, which will underpin the strategy. It is recognized that there is a need for consistency between the Acidification Strategy (which will include some consideration of eutrophication) and the Ozone Strategy, as they both involve nitrogen oxides. Work has shown that for some parts of Europe, a strategy to decrease O_3 concentrations may not be the best to reduce acidification; this problem has also to be resolved for the LRTAP Convention multi-pollutant protocol.

Effects-based approaches have not, as yet, featured in North American agreements on pollution control. Much of the early work on critical loads evolved from Canada and more recently models integrating transport and effects have been developed to target the control of emissions from particular areas. As noted above, Canadian suggestions for including critical loads in bilateral agreements have not been accepted by the USA. Meanwhile Canada has signed and ratified the 1994 Oslo Protocol indicating its agreement to the principles on which it is based and to emission reductions for the year 2000 (even though it falls outside the EMEP area for which the critical loads-based emission ceiling obligations for European countries were estimated). A recent report to the Canadian government has indicated that further major cuts in SO_2 emissions are needed if deposition is to fall below critical loads (Spurgeon, 1997). Current plans for decreasing sulphur emissions will leave large areas of Canada at risk from continued acidification. The report recommends negotiating with the USA for further reductions and for reviewing the related science, research and monitoring programmes.

Other regions across the globe are beginning to appreciate the need for emission controls and have recognized the progress made in Europe towards targeting emission reductions in a cost-effective way to achieve a certain level of environmental benefits. Critical loads maps have been prepared for other parts of the world (e.g. for Asia), and integrated assessment models have been developed for Asia (e.g. RAINSAsia) using currently available data. The UN body responsible for that area (the equivalent of the UNECE) is drawing upon the European experiences gained from the development of the LRTAP Convention. Discussions are also taking place in South America which have involved the secretariat of the LRTAP Convention from the UNECE in Geneva. In view of the rapidly increasing emissions from some parts of the world, these are important developments which are both timely and necessary to prevent wide scale problems regionally and even globally.

4   Future possibilities for emission controls

International agreements are inevitably linked with national and local pollution controls. Effects on the environment result from both local and long-range pollution transport. There is a need, therefore, to consider effects at a local scale to ensure that decreases in impacts predicted at the regional level can be realized at the local level. Often, models operating at the international scale do not match local predictions; this is inevitable given the generalization needed to derive regional strategies. Comparisons between the scales are therefore an important aspect of the work of the international community to ensure that international agreements are following their predicted trends in benefits. The monitoring and implementation of existing agreements and the compliance of parties to their defined obligations will become an important feature of LRTAP Convention protocols in the future. Increasingly, questions will be asked as to how we take further steps towards pollution control.

As ever-increasing funds are spent introducing technology to limit long-range pollution transport, a number of questions become more important. Is money being spent effectively? Is recovery possible and will benefits materialize? Are we sufficiently certain of our science and input parameters to calculate any further emission reductions required? These questions need to be considered in relation to *current reduction plans* and what are estimated to be the *maximum feasible reductions*, which may, for some countries, have an unacceptably high cost. These projected emissions from these two scenarios are illustrated in Fig. 12.1 for sulphur. Similar scenarios could be shown for nitrogen. They show that there is scope for further emission reductions of sulphur and nitrogen.

Because the costs are likely to be high, the questions posed above are important. But these questions are not answered quickly and easily. Much scientific work remains to be done and international data sets needed for definitive answers are often scant. There is controversy in some areas with regard to what is causing effects. For forests, pollution impacts are confounded with similar effects resulting from other stressors such as drought and pests. Indeed, the different stressors may act in combination or may be predisposing to one another, e.g. pollution stress may encourage pest infestation. Even the mechanism of effects resulting from soil acidification has been questioned (Løkke et al., 1996), and there are plans to investigate further methodologies for estimating critical loads. Of particular importance is the question of how long sensitive ecosystems can tolerate excess deposition, and how fast they can recover when pollution loads are reduced. For this, the steady state approaches generally used for critical loads estimates need to be complemented with dynamic assessments. As yet dynamic models have not played a major role in the development of abatement strategies; they have been limited by the availability of data and the difficulties in applying the results to strategy development. In the future such models may provide much needed insight into the discussions on the choice of effective emission reduction scenarios.

With regard to costs, the situation is also more complex than abatement negotiations might suggest. Costs used in integrated assessment models are generally for 'end of pipe' technology and may be an over-estimate of the real costs of decreasing emission by other measures, e.g. structural changes and energy conservation. In addition, measures to decrease one pollutant might also reduce others with no additional cost. The benefits are also difficult to estimate, especially in monetary terms. Consequently, cost–benefit analysis has not been applied directly for the development of abatement strategies.

However, while international agreements are being based on a simple interpretation of both science and costs, in some regions of Europe some eco-systems are showing signs of recovery from acidification. Freshwaters being monitored by the International Cooperative Programme (ICP) on assessment and monitoring of acidification of rivers and lakes, which is a subsidiary body of the LRTAP Convention, have showed in recent years small but clear signs of chemical recovery in many countries across Europe (Lükewille et al., 1997). This followed a period in the 1980s when the situation appeared to be worsening. Improvements in invertebrate assemblages have also been observed in some Norwegian and German waters. As yet improvements to soils have not been reported, but these may not be so easy to identify and for many soils the recovery time may be longer.

However, for many areas of Europe it is evident from critical loads calculations that values for critical loads are very low. It has been questioned whether it is possible to achieve critical loads and whether recovery is indeed likely. Non-linearity in the changes of deposition with respect to cuts in emissions (RGAR, 1997) suggest improvements may not be as good as most models predict. Furthermore, emissions outside of Europe and North America are continuing to increase. While these may contribute only small amounts to Europe through very long-range transport, as European emissions fall, they may increasingly contribute to the overall pollutant loads and concentrations in Europe. Estimates using global scale models, while lacking spatial resolution, suggest that meeting the ultimate targets of all critical loads and levels across Europe, as proposed by the EC, could be unrealistic for many areas if controls are only considered for Europe (R.G. Derwent, personal communication). For ozone in particular, measures being planned to reduce concentrations in Europe are likely to be offset partly by global increases. Thus there may be limits to the effectiveness of European control strategies while global agreements are likely to be much more difficult to negotiate. As noted above, there is interest in some regions of the world to adopt a similar approach to the UNECE for controlling regional emissions, for example in Asia. However, such regional agreements are only in the early stages of discussion and are unlikely to have any significant effects in the short term. Nevertheless, these will be important in the longer term to prevent significant problems across the globe.

So what can be done if critical loads cannot be achieved? Alternative approaches might use land management to reduce possible impacts. Consideration of the influence that different plant types have on the deposition of acidic pollutants (e.g. forests generally increase deposition), the sensitivity of soils to acidification and to leakage of nitrogen, and the sensitivity of plants to soil conditions, could enable planting policies which may reduce future impacts of deposited sulphur and nitrogen.

But what of areas already impacted? Large-scale liming in Scandinavia and elsewhere has shown some benefits especially in the improvement of water quality (cf. Chapter 11, this volume). Commercial agriculture and forestry might also benefit from lime or other fertilizer, though the costs of this may be prohibitive in many instances. However, for natural ecosystems such drastic measures may present problems, recovery to the existing state may be impossible and rapid alteration of soil or water conditions may influence the recovery itself. In these situations we may be faced with difficult decisions regarding expensive management plans to bring about the required changes, or a re-assessment of the value of particular areas previously protected.

Acknowledgements

KB wishes to acknowledge the UK Department of the Environment, Transport and the Regions for funding this work (contracts EPG1/3/67 and EPG1/3/116).

References

Bull, K.R. (1995) Critical loads – possibilities and constraints. *Water, Air, and Soil Pollution* 85, 201–212.

Economic Commission for Europe (ECE) (1995) *Strategies and Policies for Air Pollution Abatement: 1994 Major Review Prepared Under the Convention on Long-range Transboundary Air Pollution*. United Nations, New York and Geneva.

Grennfelt, P., Høv, O. and Derwent, R.G. (1994) Second edition abatement strategies for NO$_x$, NH$_3$, SO$_2$ and VOCs. *Ambio* 23, 425–433.

Hordijk, H. and Kroeze, C. (1997) Integrated assessment models for acid rain. *European Journal for Operational Research* 102, 405–417.

Hordijk, L. (1995) Integrated assessment models as a basis for air pollution negotiations. *Water, Air, and Soil Pollution* 85, 249–260.

Løkke, H., Bak, J., Falkengren-Grerup, U., Finlay, R.D., Ilvesniemi, H., Nygaard, P.H. and Starr, M. (1996) Critical loads of acidic deposition for forest soils: is the current approach adequate? *Ambio* 25, 510–516.

Lükewille, A., Jefferies, D., Johannessen, M., Raddum, G.G., Stoddard, J. and Traaen, T.S. (1997) *The Nine Year report. Acidification of Surface Waters in Europe and North America. Long-term Development (1980s and 1990s)*. Programme Centre, NIVA, Oslo.

Review Group on Acid Rain (RGAR) (1997) *Acid Deposition in the United Kingdom 1992–94. Fourth Report of the Review Group on Acid Rain*. AEA Technology, Abingdon.

Spurgeon, D. (1997) Canada still has a long way to go in effective control of acid rain. *Nature* 390, 8.

UNECE (1996) *1979 Convention on Long-range Transboundary Air Pollution*. United Nations, New York and Geneva.

The Future for Forests in Heavily Polluted Regions

13

J.L. Innes[1] and J. Oleksyn[2]

[1]Department of Forest Resources Management, University of British Columbia, Vancouver, Canada; [2] Polish Academy of Sciences, Institute of Dendrology, Kórnik, Poland

Air pollution has caused the degradation or loss of forests in several heavily polluted regions. In North America and Western Europe, badly affected areas are now being reclaimed. In Eastern Europe and Russia, problems continue, and affected areas are increasing. The experience gained from reclamation work in Europe and North America could usefully be applied to other regions. The greatest limitation is the availability of financial resources, with the restoration of degraded forests having low priority in many areas. Priorities need to be set for reclaiming the most affected areas. Effects-based approaches, particularly critical loads and levels, represent a suitable means by which the benefits to be gained from emission reductions can be maximized. In future, ecological risk analyses need to take into account all the possible anthropogenic impacts on forests, and should not consider pollution in isolation.

1   Introduction

The contributions to this book indicate that air pollution is still a major problem for forests in many parts of the world. For example, around Noril'sk in the Russian Federation, as much as 7 million hectares of forest may be adversely affected by pollution. Concerns about air pollution in heavily polluted industrial regions have focused on human health, but effects on forests may also be significant and, in some cases, catastrophic. In the countries of Western Europe and North America, there are few major sources of pollution still in operation; in the vast majority of cases, steps have been taken to reduce potentially damaging emissions from particular sources. Regional air pollution from large numbers of smaller sources has also been reduced. However, in Eastern Europe, Russia and Asia, pollution from point-sources remains an important problem. In such cases, it is clear that given sufficient resources, the experience gained from remedial measures in North America (e.g. Gunn, 1995; Chapter 7, this volume) and Western Europe (e.g. Chadwick and Goodman, 1975; Wilson, 1985; Moffat and McNeill, 1994) can be applied to resolving the problems in other areas.

The effects are not as clear as might be immediately assumed, with many sensationalist pictures actually showing damage associated with clear-felling of forests for fuel. However, the studies around major sources such as Sudbury and Trail in Canada and Severonikel and Noril'sk in the Russian Federation clearly indicate the importance of heavy metals and gaseous pollutants such as sulphur dioxide. The most immediate effect is the complete destruction of the forest ecosystem close to the source of the pollution. However, effects extend over much larger areas, becoming progressively more subtle with increasing distance from the source.

In some areas, the source of the pollution is not a single industrial plant. Instead, it comprises a series of smaller sources. This is the case in California (Chapter 9, this volume) and the Mediterranean Basin (Chapter 10, this volume), and is also applicable to many other areas (cf. Sandermann et al., 1997). In such circumstances, the pollution may be much more difficult to control. The debate about carbon dioxide emissions has shown, however, that if there is sufficient concern, such problems can be addressed. The main issue in such cases normally concerns the allocation of the costs of any emissions reduction programme.

Much progress has been made in the development of international agreements which help to limit pollutant emissions. These aim to stop the pollution from one country having an adverse effect on the forests of another. Such problems can only be identified and quantified through collaboration between scientists from different backgrounds, and the 1980s and 1990s saw the emergence of a number of large, interdisciplinary research projects. One example is the 1989–1994 Lapland Damage Project, which had participants from four universities and five research institutes in Finland (Tikkanen and Niemelä, 1995). In contrast with the expected findings, it concluded that the emissions from the Kola Peninsula were not expected to cause significant damage to forests in Finland. Another example is provided by the ongoing studies in Niepołomice Forest in Poland (Grodziński et al., 1984). Such projects are, however, still too rare, particularly in the Russian Federation, and better cooperation between scientists working on similar problems needs to be a high priority (Alexeyev, 1993).

The use of international agreements seems an appropriate way to control long-range transboundary air pollution. Such agreements have had success within Europe and, more recently, at the global level with the 1985 'Vienna Convention for the Protection of the Ozone Layer' and the 1987 'Montreal Protocol on Substances that Deplete the Ozone Layer'. Of particular note, the Working Group on Effects within the Convention on Long-Range Transboundary Air Pollution has had considerable success in coordinating research through a series of International Cooperative Programmes (ICPs). The largest ICP concerns the 'Assessment and Monitoring of Air Pollution Effects on Forests'. This has been very successful at mobilizing support from within Europe, with most European countries participating in the annual assessments of crown defoliation. However, it has met with rather less success in relation to identifying and monitoring pollution impacts on forests (Müller-Edzards *et al.*, 1997). In contrast, the ICP on 'Integrated Monitoring of Air Pollution Effects on Ecosystems' is much smaller, but has been more successful in relating its work to pollution impacts (Forsius *et al.*, 1998).

Air pollution in heavily industrialized regions is primarily about sulphur dioxide, the oxides of nitrogen and heavy metals. Ozone has received less attention, but Chapter 9 by Miller and Arbaugh on California and Chapter 10 by Sanz and Millán on the Mediterranean area indicate that the relative importance of this pollutant is set to increase. A major issue here is the significance of reported effects. Most work on ozone *in forests* is related to the identification of visible symptoms of injury (e.g. Skelly *et al.*, 1998). There is very little information on the impacts of such injury on the forest ecosystem as a whole, with the exception of the case study of the San Bernardino Mountains in California, USA, which has provided many useful insights into such effects (Miller and McBride, 1999).

Persistent organic pollutants (POPs) have not been covered in this report, but they are known to be important, and a legally binding Protocol to the Convention on Long-Range Transboundary Air Pollution dealing with 16 different POPs (aldrin, chlordane, dieldrin, DDT, endrin, heptachlor, hexachlorobenzene, mirex, toxaphene, PCBs, dioxins, furans, chlordecone, γ-hexachlorocyclohexane (lindane), hexabromobiphenyl, polyaromatic hydrocarbons) was agreed in June 1998. Currently, negotiations are underway for a global agreement on the first 12 of the above list. Ironically, most people's awareness of pollution problems in general probably stems from the widespread impacts of a book about POPs (Carson, 1962). As with other pollutants, a critical limits approach is being adopted to the control of POP emissions (Anon., 1998).

2   Critical loads and levels

Several authors in this volume specify critical levels (for gases) or loads (for deposited pollutants). The use of these is a step forward, but it is worth reiterating the original definition of critical loads, namely 'the highest (largest) load (of S and N pollutants) that will not cause chemical changes leading to long-term harmful effects on the most sensitive ecological systems' (Nilsson and Grennfelt, 1988). However, as it may not be practical to protect fully the most sensitive ecosystems, the definition for critical concentrations is generally worded as 'A quantitative estimate of an exposure to one or more pollutants below which significant harmful effects on specified sensitive elements of the environment do not occur according to present knowledge' (UN-ECE, 1988).

Despite the rather academic problems of definition, and the very real problems of specifying a critical load or level for a specific sensitive element, the move towards a pollution control policy based on observed effects is extremely valuable (Lövblad *et al.*, 1995). The application of the approach is dependent on the provision of suitable data. These data are often missing, and forest scientists have a valuable role to play in obtaining them. Dry deposition and occult deposition to forests are particularly important areas about which little is known. Considerable efforts are being made to specify physiologically meaningful critical levels for ozone. At present, the critical level in Europe is based on a cumulative exposure during the growing season. For forests, a critical level of 10 ppmh over 40 ppb, accumulated over a 6-month growing season, using daylight hours, is used (Skärby, 1996). However, this is of little value for predicting ozone impacts as it fails to take into account a number of mechanisms that affect the response of trees to ozone exposures. Work is currently underway to try to resolve this problem.

3   Recommendations

3.1   Open communication

One of the problems experienced when dealing with air pollution problems is that access to appropriate information has been restricted. For example, until 1988, data on environmental pollution in the former USSR were secret (Alexeyev, 1993) and only became available for the first time in 1989 (Anon., 1989). Individual polluters are naturally reluctant to provide information on their emissions, particularly peak concentrations arising from accidental emissions, but there should be no such reserve on the part of governments. The matrices produced for the Convention on Long-Range Transboundary Air Pollution, indicating the sulphur fluxes between countries, is an example of the benefits to be gained from the open exchange of information. Such information exchange should be encouraged, both within and between countries.

3.2   Integrated management and pollution problems

The considerable success that has been experienced in the 'greening up' of the Sudbury area suggests that other sites could also be improved. A major priority appears to be Noril'sk, but the smelters in the Kola Peninsula, and a number of areas in Poland, the Czech Republic and elsewhere are also important. Heale (1995) lists some of the environmental challenges and research needs facing any such programme:

- methods to prevent and control acid rock drainage;
- characterization of large waste rock piles;
- rehabilitation techniques for waste materials, including tailings disposal areas, waste rock and slag piles;
- water use, reuse, treatment and conservation management, including the impact of industrial effluent on naturally occurring wetlands;
- impact of industrial activities on groundwater movement, quality and quantity, and methods for mitigation;
- decommissioning of abandoned sites;
- impact and fate of trace metals in the environment;
- potential for biotechnology and genetic adaptation to enhance rehabilitation and restoration techniques;
- further initiatives for emission reductions and metals recovery;
- recycling and waste reduction;
- potential for ecosystem recovery with no direct treatment;
- maintenance and monitoring of perpetual treatment systems.

These should form the basis for the integrated management of heavily polluted sites.

3.3 Need for cooperation

Cooperation in research is something that is central to the aims of IUFRO. Within the context of this book, a number of issues requiring attention stand out. There is a need for the more precise use of terms and standardization of definitions. Many terminologies already exist, and IUFRO continues to be very active in this field. However, the results of these exercises need to be brought into practice. In this respect, standardization of measurement units has long been recommended, but it is surprising how many different systems are still in use.

3.4 Ecological risk analysis

Ecological risk analysis provides one mechanism by which the impacts of pollution can be assessed. While it is used extensively when looking at priorities for detoxifying polluted sites (as, for example, in the US Super Fund programme), it has not been widely used to assess the impacts of pollutants such as ozone. An exception is the work of Hogsett *et al.* (1997), which used a

GIS-based risk characterization to look at potential ozone impacts on forests in the eastern USA. The recent approach adopted by the US Environmental Protection Agency to ecological risk analysis seems particularly useful as it considers all the risks posed by anthropogenic activities, rather than the stresses created by toxic chemicals in isolation (Anon., 1991).

3.5   Air pollution and sustainable forest management

Sustainable forest management requires the maintenance of forest ecosystem health. Air pollution is a clear threat to this in heavily polluted regions. Foresters can do little to control air pollution, but they can draw attention to the extent of any problem. Forests heavily impacted by air pollution are not sustainable, and it is unlikely that wood from such forests will be certified as coming from a sustainable source. The forests themselves are unlikely to be certified as being sustainably managed, and it may be difficult for any country or region with such forests to obtain certification for all forests in the country or region.

References

Alexeyev, V.A. (1993) Air pollution impact on forest ecosystems of Kola Peninsula: history on investigations, progress and shortcomings. In: Kozlov, M.V., Haukioja, E. and Yarmishko, V.T. (eds) *Aerial Pollution in Kola Peninsula: Proceedings of the International Workshop, 14–16 April 1992, St Petersburg.* Kola Science Centre, Apatity, pp. 20–34.

Anonymous (1989) *Report on the State of the Environment in the USSR 1998.* USSR State Committee for the Protection of Nature, Moscow.

Anonymous (1991) *Summary Report on Issues in Ecological Risk Assessment.* Report no. EPA/625/3-91/018, US Environmental Protection Agency, Washington, DC.

Anonymous (1998) *Workshop on Critical Limits and Effect Based Approaches for Heavy Metals and Persistent Organic Pollutants, Bad Harzburg, Germany, 3–7 November 1997.* Report 5-98, Federal Environment Agency, Berlin.

Carson, R. (1962) *Silent Spring.* Houghton Mifflin, Boston.

Chadwick, M.J. and Goodman, G.T. (eds) (1975) *The Ecology of Resource Degradation and Renewal.* Blackwell Scientific Publications, Oxford.

Forsius, M., Guardans, R., Jenkins, A., Lundin, L. and Nielsen, K.E. (eds) (1998) *Integrated Monitoring: Environmental Assessment Through Model and Empirical Analysis. Final Results from an EU/LIFE-Project.* Finnish Environment Institute, Helsinki.

Grodziński, W., Weiner, J. and Maycock, P.F. (eds) (1984) *Forest Ecosystems in Industrial Regions. Studies on the Cycling of Energy and Pollutants in the Niepołomice Forest, Southern Poland.* Springer-Verlag, Berlin.

Gunn, J.M. (ed.) (1995) *Restoration and Recovery of an Industrial Region. Progress in Restoring the Smelter-Damaged Landscape Near Sudbury, Canada.* Springer-Verlag, New York.

Heale, E.L. (1995) Integrated management and progressive rehabilitation of industrial lands. In: Gunn, J.M. (ed.) *Restoration and Recovery of an Industrial Region. Progress in Restoring the Smelter-Damaged Landscape Near Sudbury, Canada*. Springer-Verlag, New York, pp. 287–298.

Hogsett, W.E., Weber, J.E., Tingey, D., Herstrom, A., Lee, E.H. and Laurence, J.A. (1997) Environmental auditing. An approach for characterizing tropospheric ozone risk to forests. *Environmental Management* 21, 105–120.

Lövblad, G., Grennfelt, P., Westling, O., Sverdrup, H. and Warfvinge, P. (1995) The use of critical load exceedances in abatement strategy planning. *Water, Air, and Soil Pollution* 85, 2431–2436.

Miller, P.R. and McBride, J.R. (1999) *Oxidant Air Pollution Impacts in the Montane Forests of Southern California. A Case Study of the San Bernardino Mountains*. Ecological Studies 134, Springer-Verlag, New York.

Moffat, A. and McNeill, J. (1994) *Reclaiming Disturbed Land for Forestry*. Forestry Commission Bulletin 110, HMSO, London.

Müller-Edzards, C., De Vries, W. and Erisman, J.W. (eds) (1997) *Ten Years of Monitoring Forest Condition in Europe. Studies on Temporal Development, Spatial Distribution and Impacts of Natural and Anthropogenic Stress Factors*. EC/UN-ECE, Brussels and Geneva.

Nilsson, J. and Grennfelt, P. (eds) (1988) *Critical Loads for Sulphur and Nitrogen*. Workshop, 1988/03/19-24, Skokloster, Sweden, Miljørapport 1988:97. Nordic Council of Ministers, Copenhagen.

Sandermann, H., Wellburn, A. and Heath, R.L. (eds) (1997) *Forest Decline and Ozone: a Comparison of Controlled Chamber and Field Experiments*. Ecological Studies no. 127, Springer Verlag, Berlin.

Skärby, L. (1996) Report from the forest trees group. In: Kärenlampi, L. and Skärby, L. (eds) *Critical Levels for Ozone in Europe: Testing and Finalizing the Concepts*. University of Kuopio, Kuopio, pp. 18–23.

Skelly, J.M., Innes, J.L., Snyder, K.R., Savage, J.E., Hug, C., Landolt, W. and Bleuler, P. (1998) Investigations of ozone induced injury in forests of southern Switzerland: field surveys and open-top chamber experiments. *Chemosphere* 36, 995–1000.

Tikkanen, E. and Niemelä, I. (eds) (1995) *Kola Peninsula Pollutants and Forest Ecosystems in Lapland. Final Report of The Lapland Forest Damage Project*. Finland's Ministry of Agriculture and Forestry and The Finnish Forest Research Institute, Rovanieni, Finland.

UN-ECE (1988) *ECE Critical Levels Workshop. Bad Harzburg Workshop. Final Report*. Umweltbundesamt, Berlin.

Wilson, K. (1985) *A Guide to the Reclamation of Mineral Workings for Forestry*. Forestry Commission Research and Development Paper 141, Forestry Commission, Edinburgh.

Index